Holt Pre-Al

Know-It Notebook™

HOLT, RINEHART AND WINSTON

A Harcourt Education Company

Orlando • **Austin** • New York • San Diego • Toronto • London

ISBN 0-03-038014-6

3 4 5 018 07 06 05

CONTENTS

USING THE *KNOW-IT NOTEBOOK*™

This *Know-It Notebook* will help you take notes, organize your thinking, and study for quizzes and tests. There are *Know-It Notes*™ pages for every lesson in your textbook. These notes will help you identify important mathematical information that you will need later. Then at the end of every chapter, a fun *Foldnotes* activity will help you remember key terms.

Know-It Notes

Vocabulary

One good note-taking practice is to keep a list of the new vocabulary.

- Use the page references or the glossary in your textbook to find each definition.
- Write each definition on the lines provided.

Additional Examples

Your textbook includes examples for each math concept taught. Additional examples in the *Know-It Notebook* help you take notes to remember how to solve different types of problems.

- Take notes as your teacher discusses each example.
- Write notes in the blank boxes to help you remember key concepts.
- Write final answers in the shaded boxes.

Try This

Complete the *Try This* problems that follow each lesson. Use these to make sure you understand the math concepts covered in the lesson.

- Write each answer in the space provided.
- Check your answers with your teacher or another student.
- Ask your teacher to help you understand any problem that you answered incorrectly.

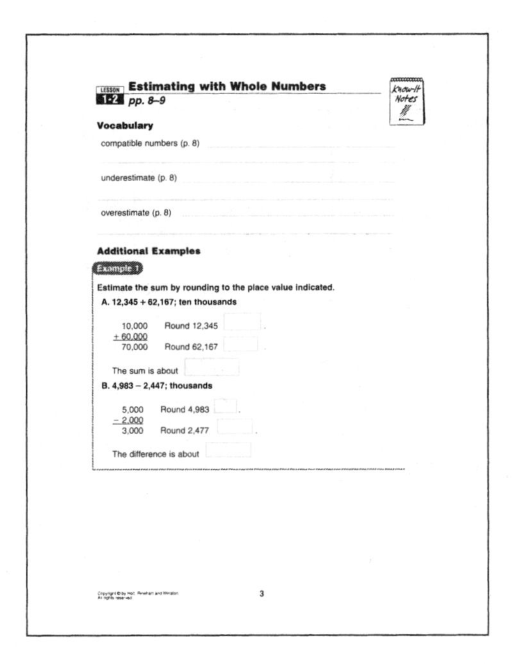
LESSON 1-2 **Estimating with Whole Numbers**
pp. 8–9

Vocabulary

compatible numbers (p. 8)

underestimate (p. 8)

overestimate (p. 8)

Additional Examples

Example 1

Estimate the sum by rounding to the place value indicated.

A. 12,345 + 62,167; ten thousands

10,000 Round 12,345
+ 60,000 Round 62,167
70,000

The sum is about

B. 4,983 − 2,447; thousands

5,000 Round 4,983
− 2,000 Round 2,477
3,000

The difference is about

3

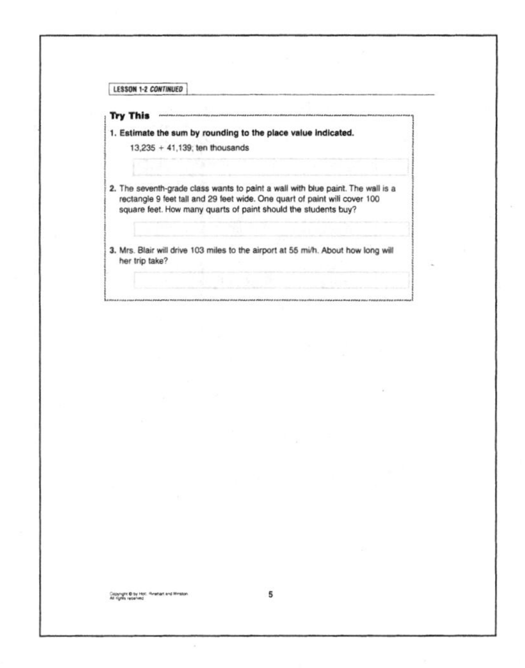
LESSON 1-2 CONTINUED

Try This

1. Estimate the sum by rounding to the place value indicated.

13,235 + 41,139; ten thousands

2. The seventh-grade class wants to paint a wall with blue paint. The wall is a rectangle 9 feet tall and 29 feet wide. One quart of paint will cover 100 square feet. How many quarts of paint should the students buy?

3. Mrs. Blair will drive 103 miles to the airport at 55 mi/h. About how long will her trip take?

5

DIRECTIONS FOR *FOLDNOTES*

Foldnotes

Each chapter ends with one of the following Foldnotes activities. Use the directions below or the directions on each activity page to make your own Foldnotes.

Vocabulary Chain

1. Write a definition and an example in words, numbers, and algebra for each term.
2. Cut out each vocabulary strip and fold in thirds.
3. Punch a hole in the corner of each folded vocabulary strip, and string them together to create your vocabulary chain.

Möbius Mobile

1. Cut each strip from the page before writing the definition.
2. Begin the definition on the same line as the word.
3. If a second line is needed, flip the strip toward you and continue on the top line. If a third line is needed, flip the strip back to the original side and continue on the next line. Continue this process until finished.
4. Hold the strip with the original side in view. Bring the two ends toward each other so the labels on the tabs are visible.
5. Flip the tab on the right and place it over tab A such that neither tab is visible.
6. Tape them in place.
7. Use string and the strips to build a Möbius mobile.

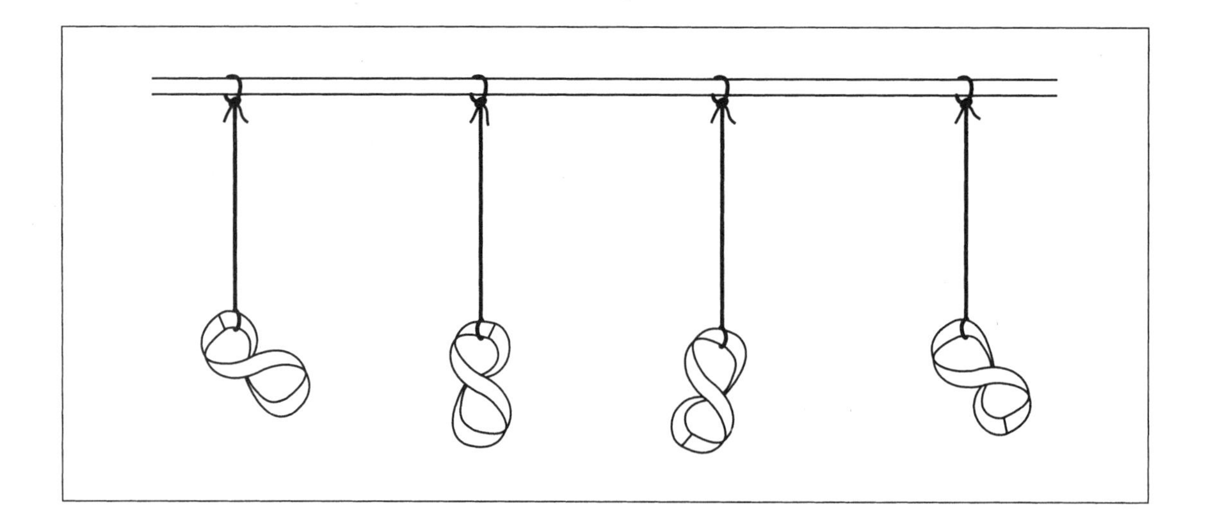

Property Prism

1. Write the definition of each term on the net of the rectangular prism.
2. Cut out the net.
3. Fold along all dotted lines, and place glue tabs to the inside of the prism.
4. Join the common edges, and tape or glue the tabs in place.

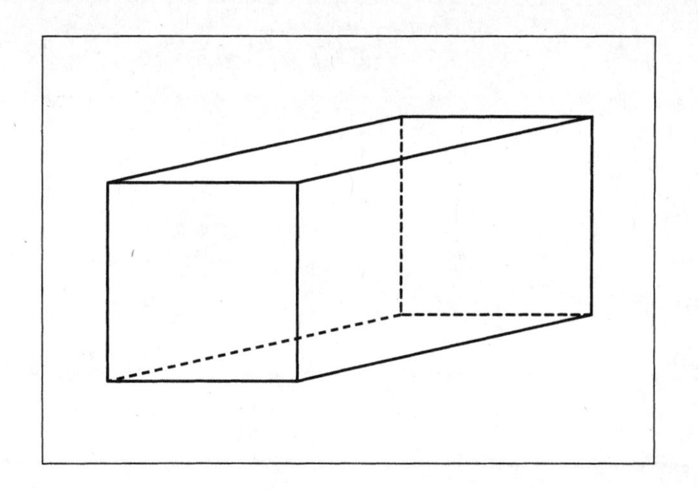

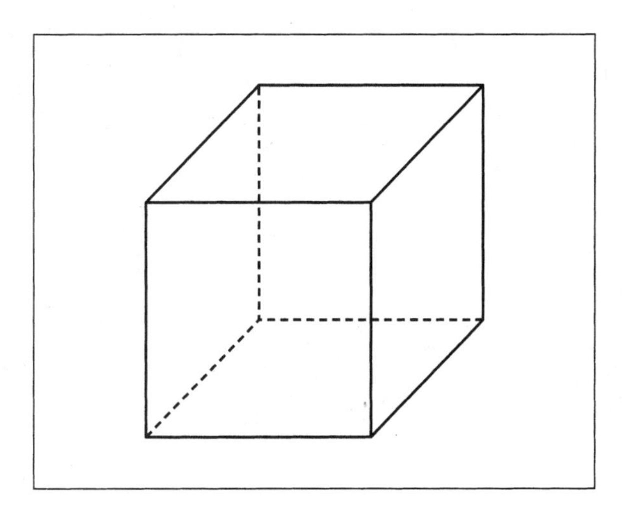

Picture Cube

1. Draw an example of each term on the net of the cube.
2. Cut out the net.
3. Fold along all dotted lines, and place glue tabs to the inside of the cube.
4. Join the common edges, and tape or glue the tabs in place.

Tetra Terms

1. Write the definition of each term on the net of the tetrahedron.
2. Cut out the net.
3. Fold along all dotted lines, and place glue tabs to the inside of the tetrahedron.
4. Join the common edges, and tape or glue the tabs in place.

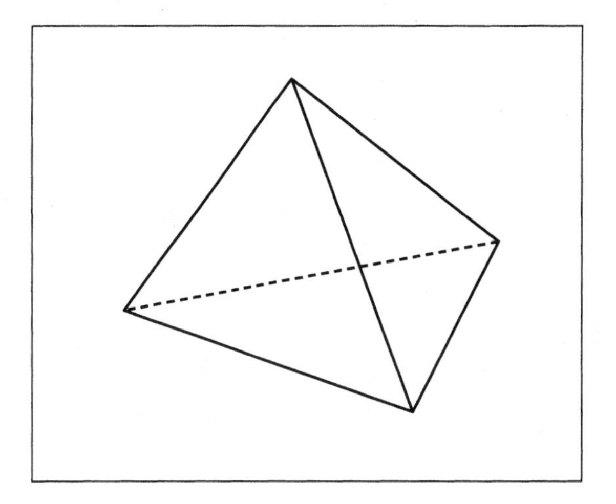

LESSON **1-1**

Variables and Expressions

pp. 4–5

Vocabulary

variable (p. 4) ____________________

____

coefficient (p. 4) ____________________

____

algebraic expression (p. 4) ____________________

____

constant (p. 4) ____________________

evaluate (p. 4) ____________________

____

substitute (p. 4) ____________________

____

Additional Examples

Example 1

Evaluate each expression for the given value of the variable.

A. $x - 5$ for $x = 12$

$\square - 5$ Substitute $\square$ for x.

$\square$ Subtract.

B. $2y + 1$ for $y = 4$

$2(\square) + 1$ Substitute $\square$ for y.

$\square + 1$ Multiply.

$\square$ Add.

Example 2

Evaluate each expression for the given value of the variables.

A. $4x + 3y$ for $x = 2$ and $y = 1$

4(   ) + 3(   )   Substitute   for x and   for y.

     +   Multiply.

     Add.

B. $8.5r - 2p$ for $r = 2$ and $p = 5.5$

8.5(   ) − 2(   )   Substitute   for r and   for p.

     −   Multiply.

     Subtract.

Try This

1. Evaluate the expression for the given value of the variable.

$4c + 1$ for $c = 11$

2. Evaluate the expression for the given value of the variables.

$8q - 3.5r$ for $q = 2.5$ and $r = 2$

LESSON **1-2**

Write Algebraic Expressions

pp. 8–10

Additional Examples

Example 1

Write an algebraic expression for each word phrase.

A. the product of 8 and a number n

the product of 8 and n

8 ☐ n

☐

B. a number w decreased by 9

w decreased by 9

w ☐ 9

☐

Example 2

A. Jared worked for h hours at the pay rate of $5 each hour. Write an expression to determine how much money Jared earned.

☐ Multiply 5 by ☐ hours.

B. How much money will he earn if he works a total of 18 hours?

☐ · 18 = ☐ Evaluate the expression for $h =$.

Jared earned $☐.

Example 3

Write an algebraic expression to evaluate each word problem.

A. Ahmed bought a new sweater for $27 plus sales tax t. If the tax was $1.76, what was the total cost of the sweater?

$ ______ + t Combine $27 with t.

$ ______ + $ ______ = $ ______ Evaluate for t = $ ______.

The total cost of the sweater was $ ______.

B. The cost to rent a banquet hall is $240. If the cost will be shared equally among all of the people who attend the event, how much will it cost each person if 12, 15, 16 or 20 people attend?

$ ______ ÷ n Separate the cost into n equal groups.

______ ÷ n In dollars

n	240 ÷ n	Cost
	240 ÷	
	240 ÷	
	240 ÷	
	240 ÷	

______ for n = 12, 15, 16, and 20.

Try This

1. A taxi-cab driver charges a base fee of $2, plus an additional $0.25 per mile. Write an expression to determine the fare.

LESSON **1-3**

Solving Equations by Adding or Subtracting *pp. 13–15*

Vocabulary

equation (p. 13)

solve (p. 13)

solution (p. 13)

inverse operation (p. 14)

isolate the variable (p. 14)

Addition Property of Equality (p. 14)

Subtraction Property of Equality (p. 14)

Additional Examples

Example 1

Determine which value of x is a solution of the equation.

$x + 8 = 15$; $x = 5, 7,$ or 23

Substitute each value for x in the equation.

$x + 8 = 15$

$5 + \square \stackrel{?}{=} 15$ Substitute $\square$ for x.

$\stackrel{?}{=} 15$ ✗

So 5 a solution.

$x + 8 = 15$

$+ 8 \stackrel{?}{=} 15$ 7 for x.

$\stackrel{?}{=} 15$ ✓

So 7 a solution.

$x + 8 = 15$

$+ 8 \stackrel{?}{=} 15$ Substitute 23 for .

$\stackrel{?}{=} 15$ ✗

So 23 a solution.

Example 2

Solve.

A. $10 + n = 18$

$10 + n = \quad 18$

$-$ ______ $-$ ______ Subtract from both sides.

$+ n =$

$n =$ Identity of Zero:

$0 + n = n$

B. $p - 8 = 9$

$p - 8 = 9$

$+ \square \quad + \square$ Add $\square$ to both sides.

$p + \square = \square$

$p = \square$ $\square$ Property of Zero: $p + 0 = p$

C. $22 = y - 11$

$22 = y - 11$

$+ \square \quad + \square$ Add $\square$ to both sides.

$\square = y + \square$

$\square = y$ Identity Property of Zero: $y + \square = \square$

Try This

1. Determine which value of *x* is a solution of the equation.

$x - 4 = 13$; $x = 9, 17$, or 27

2. Solve.

$44 = y - 23$

LESSON **1-4**

Solving Equations by Multiplying or Dividing *pp. 18–19*

Vocabulary

Division Property of Equality (p. 18)

Multiplication Property of Equality (p. 19)

Additional Examples

Example 1

Solve $8x = 32$.

$8x = 32$

$\frac{8x}{} = \frac{32}{}$ Divide both sides by .

$1x =$ $1 \cdot x =$

$x =$

Check

$8x = 32$

$8(\quad) \overset{?}{=} 32$ Substitute for x.

$\overset{?}{=} 32$ ✓

Example 2

Solve $\frac{n}{7} = 7$.

$\square \cdot \frac{n}{7} = \square \cdot 7$ Multiply both sides by $\square$.

$n = \square$

Example 5

Solve $3y - 7 = 20$.

Step 1: $3y - 7 = 20$

$+\square \quad +\square$ Add $\square$ to both sides to $\square$ the term with y in it.

$3y = \square$

Step 2: $\frac{3y}{\square} = \frac{27}{\square}$ Divide both sides by $\square$.

$y = \square$

Try This

1. Solve $9x = 36$.

2. Solve $\frac{n}{4} = 16$.

Solving Simple Inequalities
pp. 23–25

Vocabulary

inequality (p. 23)

algebraic inequality (p. 23)

solution of an inequality (p. 23)

solution set (p. 23)

Additional Examples

Example 1

Compare. Write $<$ or $>$.

A. $23 - 14$ ▢ 6

6

B. $5(12)$ ▢ 70

70

Example 2

Solve and graph the inequality.

$x + 2.5 \leq 8$

$-$ ______ $-$ ______ Subtract from both sides.

$x \leq$

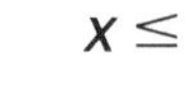

1 2 3 4 5 6 7

Example 3

PROBLEM SOLVING APPLICATION

An interior designer is planning to place a wallpaper border along the edges of all four walls of a room. The total distance around the room is 88 feet. The border comes in packages of 16 feet. What is the least number of packages that must be purchased to be sure that there is enough border to complete the room?

1. Understand the Problem

The answer will be the least number of packages of border needed to wallpaper a room.

List the important information:

- The total distance around the room is ______ feet.
- The border comes in packages of ______ feet.

Show the relationship of the information:

the number of packages of border	•	the length of one package of border	≥	88 feet

2. Make a Plan

Use the relationship to write an ______. Let x represent the number of packages of border.

x	•	16 ft	≥	88 feet

3. Solve

______ ≥ ______

______ ≥ ______ Divide both sides by ______.

$x \geq$ ______

At least ______ packages of border must be used to complete the room.

4. Look Back

Because whole packages of border must be purchased, at least packages of border must be purchased to ensure that there is enough to complete the room.

Try This

1. Compare. Write $<$ or $>$.

4(15) ___ 50

2. Solve and graph the inequality.

$6u > 72$

3 6 9 12 15 18 21

3. Problem Solving Application

Ron will provide 130 cookies for the school fundraiser. He has to buy the cookies in packages of 20. What is the least number of packages Ron must buy to be sure to have enough cookies?

LESSON **1-6**

Combining Like Terms

pp. 28–29

Vocabulary

term (p. 28) The parts of an expression that are or subtracted

like term (p. 28) TWO or more terms that have the same variable raised to the same power

equivalent expression (p. 28) Having the same value

simplify (p. 29) To write a fraction or expression In simplest form.

Additional Examples

Example 1

Combine like terms.

A. $14a - 5a$

Identify ______ terms.

Combine coefficients: ____ $-$ ____ $=$ ____

B. $7y + 8 - 3y - 1 + y$

Identify like ______; the ______ of y is 1, because $1y = y$.

Combine ______:

$7 - 3 + 1 = 5$ and $8 - 1 = 7$

Example 2

Combine like terms.

$9x + 3y - 2x + 5$

$9x + 3y - 2x + 5$ Identify terms.

Combine coefficients: $- \quad = 7$

Example 3

Simplify $6(5 + n) - 2n$.

$6(5 + n) - 2n$

$6(5) + 6(n) - 2n$ Property

$+ 6n - 2n$ $6n$ and $2n$ are terms.

$+ \quad n$ Combine coefficients: $6 - 2 =$

Try This

1. Combine like terms.

$5c + 8 - 4c - 2 - c$

2. Combine like terms.

$9d + 7c - 4d - 2c$

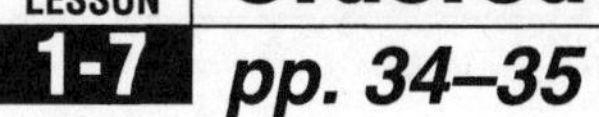

Ordered Pairs

pp. 34–35

Vocabulary

ordered pair (p. 34) ______________________

____

Additional Examples

Example 1

Determine whether the ordered pair is a solution of $y = 4x - 1$.

A. (3, 11)

$y = 4x - 1$

$\_\_\_ \stackrel{?}{=} 4(\_\_\_) - 1$ Substitute ___ for x and ___ for y.

$\_\_\_ \stackrel{?}{=} \_\_\_$ ✓ A ______________ since 11 = 11

(3, 11) ___ a solution.

B. (10, 3)

$y = 4x - 1$

$\_\_\_ \stackrel{?}{=} 4(\_\_\_) - 1$ Substitute 10 for ___ and 3 for ___.

$\_\_\_ \stackrel{?}{=} \_\_\_$ ✗

(10, 3) ______ a solution.

Example 2

Use the given values to make a table of solutions.

$y = 7x$ for $x = 1, 2, 3, 4$

x	$7x$	y	(x, y)
1			(,)
2			(,)
3			(,)
4			(,)

Try This

1. Determine whether the ordered pair is a solution of $y = 5x + 3$.

(9, 17)

2. Use the given values to make a table of solutions.

$y = 6x$ for $x = 1, 2, 3, 4$

x	$6x$	y	(x, y)
1			(,)
2			(,)
3			(,)
4			(,)

LESSON 1-8

Graphing on a Coordinate Plane

pp. 38–39

Vocabulary

coordinate plane (p. 38)

x-axis (p. 38)

y-axis (p. 38)

x-coordinate (p. 38)

y-coordinate (p. 38)

origin (p. 38)

graph of an equation (p. 39)

Additional Examples

Example 1

Give the coordinates of each point.

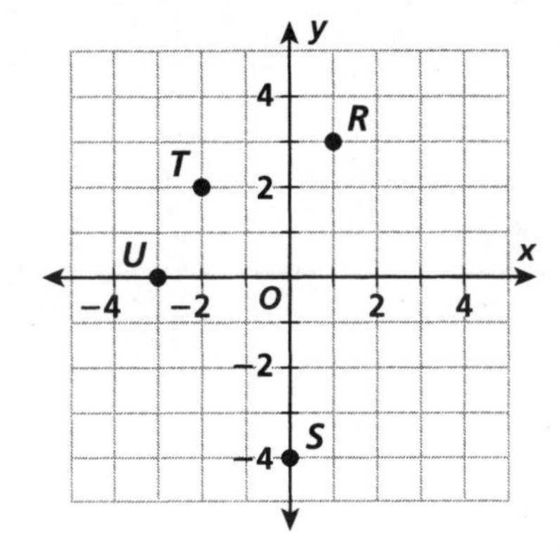

Point *R* is (,).

unit right, units up

Point *S* is (,).

units right, units down

Point T is (,) .

2 units , 2 units

Point U is (,) .

3 units , 0 units

Try This

1. Give the coordinates of each point.

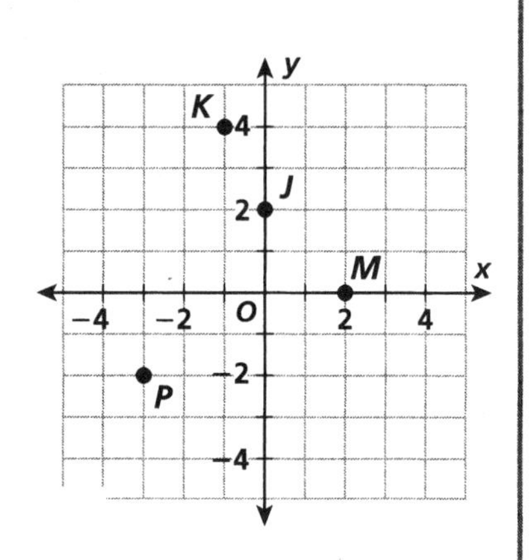

(,), (,), (,), (,)

2. Graph each point on the coordinate plane. Label the points *A–D*.

$A(2, 4)$, $B(2, 0)$, $C(-2, 4)$, $D(-1, -1)$

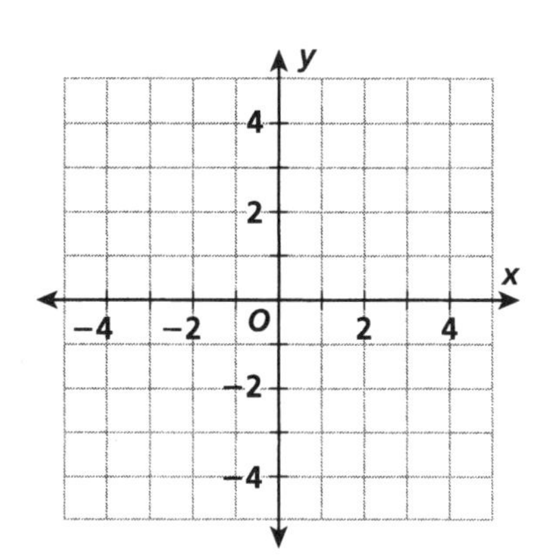

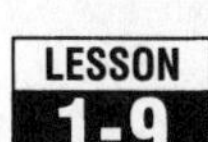

Interpreting Graphs and Tables

pp. 43–44

Additional Examples

Example 1

The table gives the speeds in mi/h of three cars at given times. Tell which car corresponds to each situation described below.

Time	1:00	1:05	1:10	1:15	1:20
Car 1	50	50	30	30	0
Car 2	55	50	50	55	50
Car 3	55	10	0	0	55

A. Mr. Lee is traveling on the highway, and he pulls over to make a call on his cell phone. Then he gets back onto the highway.

Car ____ —Mr. Lee starts to slow down after ________. After making his phone call, he gets back onto the highway and resumes his speed of ____ mi/h.

B. Mrs. Healy is driving on the freeway between 1:00 and 1:20 and encounters no traffic.

Car ____ —Mrs. Healy's car maintains a speed between ____ mi/h and ____ mi/h.

C. At 1:00 Mr. Johns estimates that he will arrive at his destination in approximately 20 minutes.

Car ____ —Mr. Johns's speed begins to decrease to 0 sometime between ________ and ________.

Example 2

Tell which graph corresponds to each situation described in Additional Example 1.

A.

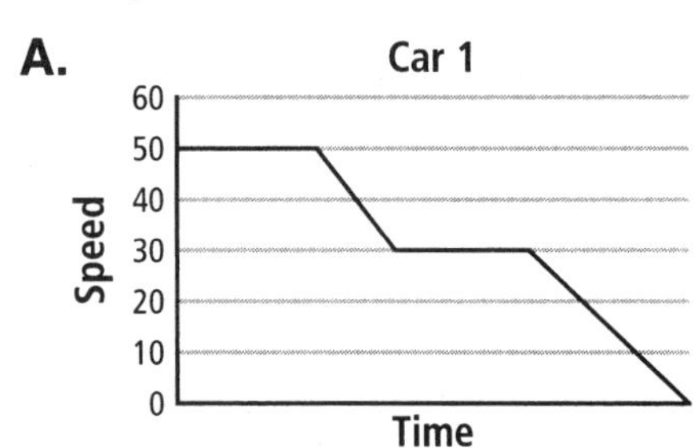

B.

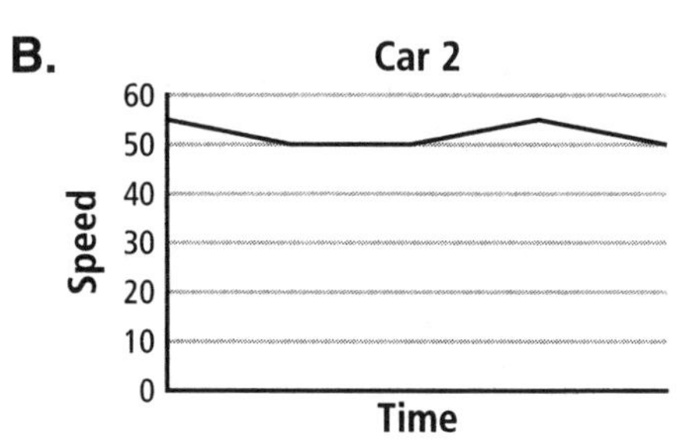

C.

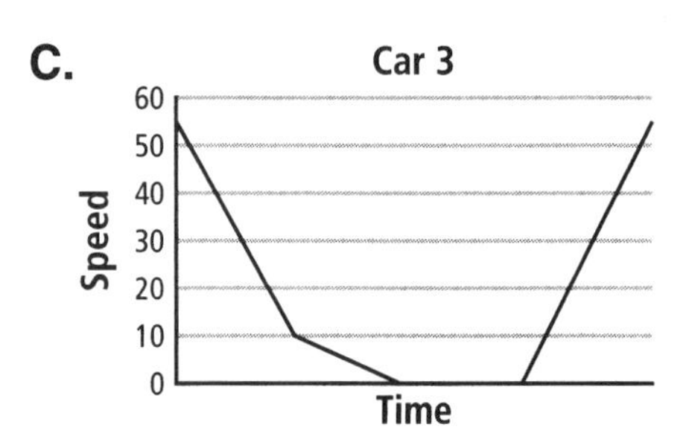

Try This

1. The table gives the speed in mi/h of three runners at the given times. Tell which runner corresponds to the situation described below.

Time	8:00	8:10	8:20	8:30	8:40
Runner 1	6	7	7	7	7
Runner 2	9	6	0	0	0
Runner 3	8	3	8	3	8

Jamie begins the race, and soon feels a pain in a muscle. He is unable to complete the race.

Chapter 1
Property Prism

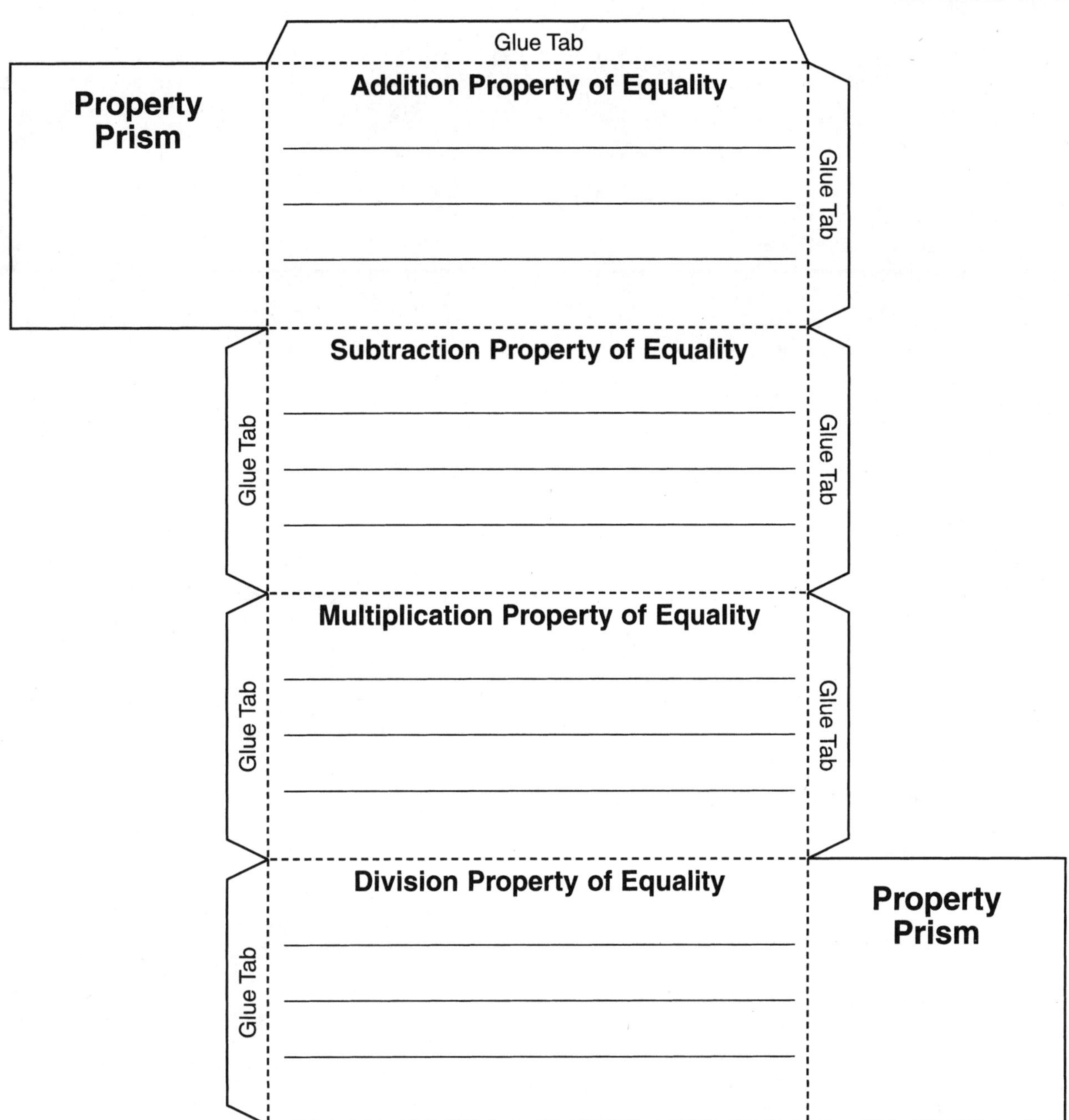

Directions

1. Write the definition of each term on the net.
2. Cut out the net of the rectangular prism.
3. Fold along all dotted lines, and place glue tabs to the inside of the prism.
4. Join the common edges, and tape or glue the tabs in place.

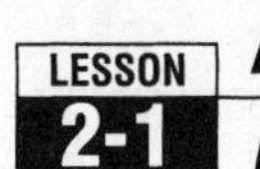

Adding Integers
pp. 60–61

Vocabulary

integers (p. 60)

opposites (p. 60)

absolute value (p. 60)

Additional Examples

Example 1

Use a number line to find the sum.

$(-6) + 2$

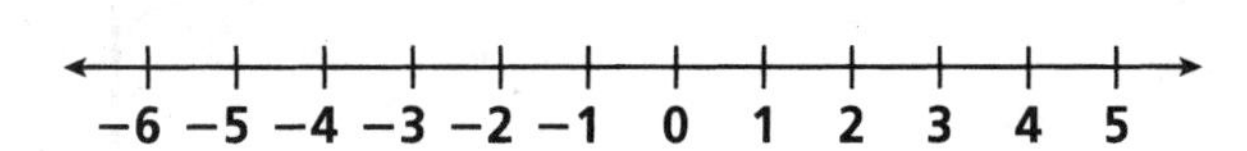

You finish at ____, so $(-6) + 2 =$ ____.

Example 2

Add.

$1 + (-2)$ Think: Find the ____ of 2 and 1.

____ $2 > 1$; use the sign of ____.

Example 3

Evaluate $c + 4$ for $c = -8$.

$c + 4$

$(\quad) + 4$ Replace c with ____.

Think: Find the ____ of 8 and 4.

$\quad + 4 =$ $8 > 4$; use the sign of ____.

Try This

1. Use a number line to find the sum.

$(-3) + 7$

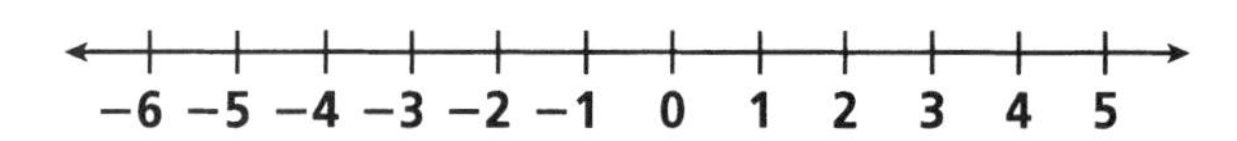

2. Add.

$(-4) + 1$

3. Evaluate $d + 7$ for $d = -3$.

LESSON 2-2

Subtracting Integers

pp. 64–65

Additional Examples

Example 1

Subtract.

A. $-7 - 4$

$-7 - 4 = -7 + (\quad)$ Add the ______ of 4.

$= \quad$ Same sign; use the sign of the integers.

B. $8 - (-5)$

$8 - (-5) = 8 \quad 5$ Add the ______ of –5.

$= \quad$ Same sign; use the sign of the ______.

C. $-6 - (-3)$

$-6 - (-3) = -6 \quad 3$ Add the opposite of ______.

$= \quad$ $6 > 3$; use the sign of ______.

Example 2

Evaluate the expression for the given value of the variable.

A. $8 - j$ for $j = -6$

$8 - j$

$8 - (\quad)$ Substitute ______ for j.

$= 8 \quad 6$ Add the ______ of -6.

$= \quad$ Same sign; use the sign of the ______.

B. $-9 - y$ for $y = -4$

$-9 - y$

$-9 - (\quad)$ Substitute -4 for ___.

$= -9 \quad 4$ Add the opposite of ___.

$=$ $9 > 4$; use the ___ of 9.

C. $n - 6$ for $n = -2$

$n - 6$

$\quad - 6$ Substitute ___ for n.

$= -2 \quad (-6)$ ___ the opposite of 6.

$=$ Same sign; use the sign of the ___.

Try This

1. Subtract.

$-7 - (-8)$

2. Evaluate the expression for the given value of the variable.

$-5 - r$ for $r = -2$

LESSON **2-3**

Multiplying and Dividing Integers

pp. 68–69

Additional Examples

Example 1

Multiply or divide.

A. $-6(4)$ Signs are ______.

______ Answer is ______.

B. $-8(-5)$ Signs are the ______.

______ Answer is ______.

C. $\frac{-18}{2}$ ______ are different.

______ Answer is ______.

D. $\frac{-25}{-5}$ ______ are the same.

______ Answer is ______.

Example 2

Simplify.

A. $3(-6 - 12)$ Subtract inside the ______.

$= 3($ ______ $)$ Think: The signs are ______.

$=$ ______ The answer is ______.

Example 3

Complete a table of solutions for $y = 3x - 1$ for $x = -2, -1, 0, 1$, and 2. Plot the points on a coordinate plane.

x	$3x - 1$	y	(x, y)
			(,)
			(,)
			(,)
			(,)
			(,)

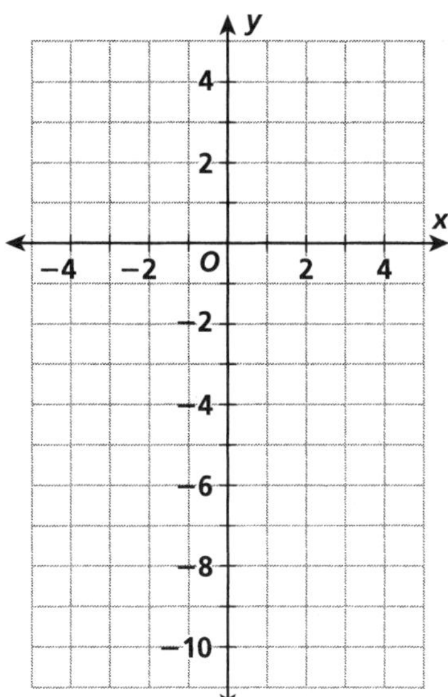

Try This

1. Multiply or divide.

$-3(-2)$

2. Simplify.

$-3(6 - 9)$

LESSON **2-4**

Solving Equations Containing Integers *pp. 74–75*

Additional Examples

Example 1

Solve.

A. $x - 3 = -6$

$x - 3 = -6$

$x - 3$ ☐ $3 = -6$ ☐ 3 ☐ 3 to both sides.

$x =$ ☐ $x \underbrace{- 3 + 3}_{0} = -3$

B. $-5 + r = 9$

$-5 + r = 9$

$-5 +$ ☐ $+ r = 9 +$ ☐ Add 5 to both sides.

$r =$ ☐

C. $-6 + 8 = n$

$-6 + 8 = n$ The ☐ is already isolated.

☐ $= n$ Add integers.

D. $z + 6 = -3$

$z + 6 = -3$

$\underline{+ (-6)}$ $\underline{+ (-6)}$ Add ☐ to each side.

$z =$ ☐

Example 2

Solve.

A. $-5x = 35$

$\quad = \quad$ Divide both sides by ___.

$x =$

B. $\frac{z}{-4} = 5$

$\cdot \frac{z}{-4} = \quad \cdot 5$ Multiply both sides by ___.

$z =$

Try This

1. Solve.

$a + 9 = -9$

2. Solve.

$\frac{z}{-3} = 9$

Solving Inequalities Containing Integers *pp. 78–79*

Additional Examples

Example 1

Solve and graph.

A. $k + 3 > -2$

$k + 3 > -2$

$-$ ___ $-$ ___ Subtract ___ from both sides.

$k >$ ___

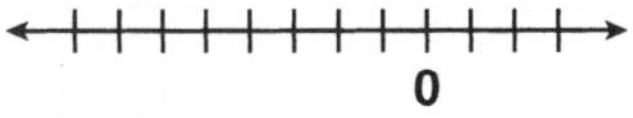

B. $r - 9 \geq 12$

$r - 9 \geq 12$

$r - 9 +$ ___ $\geq 12 +$ ___ Add ___ to both sides.

$r \geq$ ___

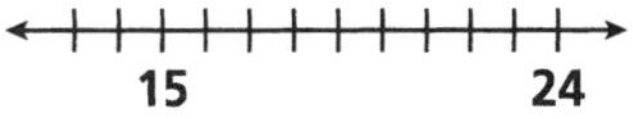

C. $u - 5 \leq 3$

$u - 5 \leq 3$

$u - 5 + 5$ ___ $3 + 5$ ___ 5 to both sides.

u ___ ___

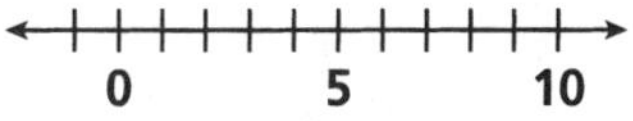

Example 2

Solve and graph.

$-3y \geq 15$

Divide each side by ; $\geq$ changes to .

y

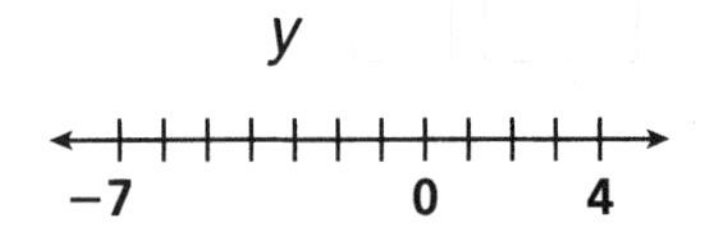

Try This

1. Solve and graph.

$y + 7 \geq -1$

2. Solve and graph.

$-8y \geq 24$

LESSON 2-6

Exponents

pp. 84–85

Vocabulary

power (p. 84) ______________________________

exponential form (p. 84) ______________________________

exponent (p. 84) ______________________________

base (p. 84) ______________________________

Additional Examples

Example 1

Write in exponential form.

A. $4 \cdot 4 \cdot 4 \cdot 4$

$4 \cdot 4 \cdot 4 \cdot 4 =$ ______ Identify how many times ______ is a factor.

B. $d \cdot d \cdot d \cdot d \cdot d$

$d \cdot d \cdot d \cdot d \cdot d =$ ______ Identify how many times d is a ______.

Example 2

Evaluate.

A. 3^5

$3^5 =$ ______ Find the product of ______ 3's.

$=$ ______

B. $(-3)^5$

$(-3)^5 =$ Find the product of

−3's.

$=$

Example 3

Simplify $(2^5 - 3^2) + 6(4)$

$= (\quad - \quad) + 6(4)$ Evaluate the .

$= (\quad) + 6(4)$ Subtract inside the .

$= 23 +$ from left to right.

$=$ from left to right.

Try This

1. Write in exponential form.

$(-3) \cdot (-3) \cdot (-3) \cdot (-3)$

2. Evaluate.

$(-9)^3$

3. Simplify $(3^2 - 8^2) + 2 \cdot 3$.

LESSON 2-7

Properties of Exponents

pp. 88–89

Know-It Notes

Additional Examples

Example 1

Multiply. Write the product as one power.

A. $6^6 \cdot 6^3$

exponents.

B. $n^5 \cdot n^7$

exponents.

Example 2

Divide. Write the product as one power.

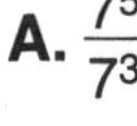

A. $\frac{7^5}{7^3}$

exponents.

B. $\frac{x^{10}}{x^9}$

Subtract .

Think: $x^1 =$

Example 3

A light-year, or the distance light travels in one year, is almost 10^{18} centimeters. To convert this number to kilometers, you must divide by 10^5. How many kilometers is a light-year?

$\frac{10^{18}}{10^5}$

Subtract .

A light-year is almost km.

Try This

1. Multiply. Write the product as one power.

$4^2 \cdot 4^4$

2. Divide. Write the product as one power.

$\frac{9^9}{9^2}$

3. A ship has 10^7 kilograms of grain loaded into its cargo hold. A metric ton is 10^3 kilograms. How many metric tons of grain were loaded?

LESSON **2-8**

Look for a Pattern in Integer Exponents *pp. 92–93*

Additional Examples

Example 1

Evaluate the powers of 10.

A. 10^{-2}

$10^{-2} = \frac{1}{10 \cdot 10}$

$10^{-2} = \square = \square$

B. 10^{-1}

$10^{-1} = \frac{1}{10}$

$10^{-1} = \square = \square$

C. 10^{-6}

$10^{-6} = \frac{1}{10 \cdot 10 \cdot 10 \cdot 10 \cdot 10 \cdot 10}$

$10^{-6} = \square = \square$

Example 2

Evaluate 5^{-3}.

$\frac{1}{5^3}$ Write the __________; change the __________ of the exponent.

$\frac{1}{5 \cdot 5 \cdot 5}$

$\square$

Example 3

Evaluate.

$2^{-5} \cdot 2^{3}$

2^{-5+3} Bases are the same, so the exponents.

Write the ; change the sign of the

$\frac{1}{2^{2}}$.

Check: $2^{-5} \cdot 2^{3} = \frac{1}{2^{5}} \cdot 2^{3} = \frac{2^{3}}{2^{5}} = \frac{2 \cdot 2 \cdot 2}{2 \cdot 2 \cdot 2 \cdot 2 \cdot 2} = \frac{1}{4}$

Try This

1. Evaluate the power of 10.

10^{-8}

2. Evaluate $(-10)^{-3}$.

3. Evaluate.

$7^{-6} \cdot 7^{7}$

LESSON **2-9** # Scientific Notation
pp. 96–97

Vocabulary

scientific notation (p. 96) ______

Additional Examples

Example 1

Write each number in standard notation.

A. 1.35×10^5

1.35×10^5

$1.35 \times$ ______ $10^5 =$ ______

______ Think: Move the decimal right ___ places.

B. 2.7×10^{-3}

2.7×10^{-3}

$2.7 \times$ ______ $10^{-3} =$ ______

2.7 ___ 1000 ______ by the reciprocal.

______ Think: Move the decimal ______ 3 places.

C. -2.01×10^4

-2.01×10^4

$-2.01 \times$ ______ $10^4 =$ ______

______ Think: Move the decimal right ___ places.

Example 2

Write 0.00709 in scientific notation.

0.00709

Move the decimal to get a number between and .

7.09×10 Set up notation.

Think: The decimal needs to move left to change 7.09 to

0.00709, so the exponent will be .

Think: The decimal needs to move places.

So 0.00709 written in scientific notation is $\times$.

Check $\times$ $= 7.09 \times 0.001 = 0.00709$

Try This

1. Write the number in standard notation.

-5.09×10^8

2. Write 0.000811 in scientific notation.

Chapter 2
Möbius Mobile

absolute value ____

base ____

exponent ____

exponential form ____

integer ____

opposite ____

power ____

scientific notation ____

Directions

1. Cut each strip from the page before writing the definition.
2. Begin the definition on the same line as the word.
3. If a second line is needed, flip the strip toward you and continue on the top line. If a third line is needed, flip the strip back to the original side and continue on the next line. Continue this process until finished.
4. Hold the strip with the original side in view. Bring the two ends toward each other so the labels on the tabs are visible.
5. Flip the tab on the right and place it over tab A such that neither tab is visible.
6. Tape them in place.
7. Use string and the strips to build a Möbius mobile.

Chapter 2
Möbius Mobile

Flip		Tab A
Flip		Tab A
Flip		Tab A
Flip		Tab A
Flip		Tab A
Flip		Tab A
Flip		Tab A
Flip		Tab A

LESSON **3-1**

Rational Numbers

pp. 112–114

Vocabulary

rational number (p. 112) ______

relatively prime (p. 112) ______

Additional Examples

Example 1

Simplify.

$\frac{5}{10}$

$5 = 1 \cdot 5$ ___ is a common factor.

$10 = 2 \cdot 5$

$\frac{5}{10} = \frac{5 \div \square}{10 \div \square}$ Divide the numerator and denominator by ___.

$= \square$

Example 2

Write the decimal as a fraction in simplest form.

-0.8

$= \frac{\square}{10}$ -8 is in the ______ place.

$= \square$ Simplify by dividing by the common factor ___.

Example 3

Write the fraction as a decimal.

$\frac{11}{9}$

The pattern repeats, so draw a bar over the 2 to indicate that

this is a decimal.

The fraction $\frac{11}{9}$ is equivalent to the decimal .

Try This

1. Simplify.

$\frac{18}{27}$

2. Write the decimal as a fraction in simplest form.

8.75

3. Write the fraction as a decimal.

$\frac{9}{40}$

LESSON **3-2**

Adding and Subtracting Rational Numbers *pp. 117–118*

Additional Examples

Example 1

In August 2001 at the World University Games in Beijing, China, Jimyria Hicks ran the 200-meter dash in 24.08 seconds. Her best time at the U.S. Senior National Meet in June of the same year was 23.35 seconds. How much faster did she run in June?

$$\begin{array}{r} 24.08 \\ -23.35 \\ \hline \end{array}$$ Align the decimals.

She ran ______ second faster in June.

Example 2

Use a number line to find the sum.

A. 0.3 + (−1.2)

Move right ______ units.

−1.4 −1.0 −0.4 0 0.4

From ______, move left ______ units.

You finish at ______, so 0.3 + (−1.2) = ______.

Example 3

Add or subtract.

$-\frac{2}{9} - \frac{5}{9}$

$-\frac{2}{9} - \frac{5}{9} = \frac{\qquad\qquad}{9} =$ Subtract numerators. Keep the denominator.

Example 4

Evaluate the expression for the given value of the variable.

$12.1 - x$ for $x = -0.1$

$12.1 - (\qquad)$ Substitute for x.

Think: $12.1 - (-0.1) = 12.1 \qquad 0.1$

Try This

1. Tom ran the 100-meter dash in 11.5 seconds last year. This year he improved his time by 0.568 seconds. How fast did Tom run the 100-meter dash this year?

2. Use a number line to find the sum.

$1.5 + (-1.8)$

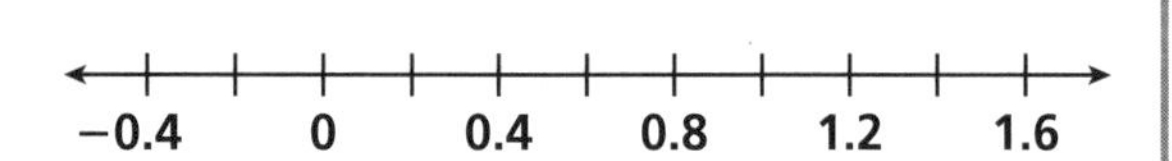

3. Add or Subtract.

$-\frac{1}{5} - \frac{3}{5}$

LESSON **3-3**

Multiplying Rational Numbers

pp. 121–123

Additional Examples

Example 1

Multiply. Write the answer in simplest form.

$-8\left(\frac{6}{7}\right)$

$= \frac{-8 \cdot 6}{7}$

$=$ ______ Multiply.

$=$ ______ Simplify.

Example 2

Multiply. Write the answer in simplest form.

$\frac{1}{8}\left(\frac{6}{7}\right)$

$= \frac{1(6)}{8(7)}$ Multiply ______.

Multiply ______.

$= \frac{1(6)}{8(7)}$ Look for common ______: 2.

$=$ ______ Simplest form

Example 3

Multiply.

$2(-0.51)$

$2 \cdot (-0.51) =$ Product is with 2 decimal places.

Try This

1. Multiply. Write the answer in simplest form.

$-3\left(\frac{5}{8}\right)$

2. Multiply. Write the answer in simplest form.

$\frac{3}{5}\left(\frac{5}{8}\right)$

3. Multiply.

$3.1(0.28)$

LESSON 3-4

Dividing Rational Numbers

pp. 126–128

Vocabulary

reciprocal (p. 126) ____________________

____

Additional Examples

Example 1

Divide. Write the answer in simplest form.

$\frac{5}{11} \div \frac{1}{2}$

$\frac{5}{11} \div \frac{1}{2} = \frac{5}{11} \cdot \square$ Multiply by the ____________.

$= \frac{5}{11} \cdot \square$ No common ____________

$= \square$ Simplest form

Example 2

Divide.

$0.384 \div 0.24$

$0.384 \div 0.24 = \frac{0.384}{0.24}\left(\frac{100}{100}\right) = \square$

$= \square$ Divide.

Example 3

Evaluate the expression for the given value of the variable.

$\frac{5.25}{n}$ for $n = 0.15$

$\frac{5.25}{0.15} = \frac{5.25}{0.15}\left(\frac{100}{100}\right)$ 0.15 has 2 decimal places, so use .

$=$ Divide.

$=$

Example 4

PROBLEM SOLVING APPLICATION

A cookie recipe calls for $\frac{1}{2}$ cup of oats. You have $\frac{3}{4}$ cup of oats. How many batches of cookies can you bake using all of the oats you have?

1. **Understand the Problem**
   The number of batches of cookies you can bake is the number of batches using the oats that you have. List the important information:

   - The amount of oats is cup.
   - One batch of cookies calls for cup of oats.

2. **Make a Plan**
   Set up an equation.

amount of oats you have	÷	amount for one batch	=	number of batches

3. **Solve**
   Let n = number of batches.

   $\frac{3}{4} \div \frac{1}{2} = n$

   $\frac{3}{4} \cdot \quad = n$

   $\frac{6}{4}$, or batches of the cookies.

4. Look Back

One cup of oats would make ______ batches so $1\frac{1}{2}$ is a ______ answer.

Try This

1. Evaluate the expression for the given value of the variable.

$\frac{2.55}{b}$ for $b = 0.75$

2. Problem Solving Application

A ship will use $\frac{1}{6}$ of its total fuel load for a typical round trip. If there is $\frac{5}{8}$ of a total fuel load on board now, how many complete trips can be made?

1. Understand the Problem

The number of complete trips the ship can make is the number of trips that the ship can make with the fuel on board. List the important information:

- It takes ______ of the total fuel load for a complete trip.
- You have ______ of a total fuel load on board right now.

2. Make a Plan

Set up an equation.

Amount of fuel on board	÷	Amount of fuel for one trip	=	Number of trips

3. Solve

4. Look Back

A full tank will make the round trip 6 times, and $\frac{5}{8}$ is a little more than $\frac{1}{2}$, so half of 6, or 3, is a ______ answer.

LESSON **3-5**

Adding and Subtracting with Unlike Denominators *pp. 131–132*

Vocabulary

least common denominator (LCD) (p. 131)

Additional Examples

Example 1

Add or Subtract.

Method 1:

A. $\frac{1}{8} + \frac{2}{7}$ Find a common :

8(7) = .

$= \frac{1}{8}(\quad) + \frac{2}{7}(\quad)$ Multiply by fractions equal to .

$=\quad +$ Rewrite with a common

.

$=$ Simplify

Method 2:

B. $1\frac{1}{6} + \frac{5}{8}$

$= \square + \frac{5}{8}$ — Write as an ______ fraction.

Multiples of 6: 6; 12; 24; 30

Multiples of 8: 8; 16; 24; 32

List the ______ of each denominator and find the ______.

$= \frac{7}{6}\square + \frac{5}{8}\square$ — Multiply by fractions equal to ______.

$= \frac{\square}{24} + \frac{\square}{24} = \frac{\square}{24}$ — Rewrite with a ______ denominator.

$= \square$ — Simplify.

Try This

1. Add.

$2\frac{1}{6} + \frac{3}{4}$

2. Evaluate $\frac{5}{9} - h$ for $h = \frac{-7}{12}$.

LESSON **3-6**

Solving Equations with Rational Numbers *pp. 136–137*

Additional Examples

Example 1

Solve.

A. $m + 4.6 = 9$

$m + 4.6 = 9$

$\underline{-\quad\quad} = \underline{-\quad\quad}$ Subtract from both sides.

$m =$

B. $8.2p = -32.8$

$\frac{8.2p}{\quad} = \frac{-32.8}{\quad}$ Divide both sides by .

$p =$

Example 2

Solve.

$n + \frac{2}{7} = -\frac{3}{7}$

$n - \quad + \frac{2}{7} = -\frac{3}{7} -$ Subtract from both sides.

$n =$

Example 3

Mr. Rios wants to prepare a casserole that requires $2\frac{1}{2}$ cups of milk. If he makes the casserole, he will have only $\frac{3}{4}$ cup of milk left for his breakfast cereal. How much milk does Mr. Rios have?

Convert fractions: $2\frac{1}{2} = \frac{2(2) + 1}{2} = \square$

Write an equation:

Original amount of milk	−	*Milk for casserole*	=	*Milk for cereal*
c	−	$\frac{5}{2}$	=	$\frac{3}{4}$

Now solve the equation.

$$c - \frac{5}{2} = \frac{3}{4}$$

$c - \frac{5}{2} + \square = \frac{3}{4} + \square$ Add $\square$ to both sides.

$c = \frac{\square}{4} + \frac{\square}{4}$ Find a $\square$ denominator, 4.

$c = \square$, or $\square$ Simplify.

Mr. Rios has $\square$ cups of milk.

Try This

Solve.

1. $5.5b = 75.9$

2. $\frac{3}{8}x = \frac{6}{19}$

LESSON 3-7

Solving Inequalities with Rational Numbers *pp. 140–141*

Additional Examples

Example 1

Solve.

$0.4x \leq 0.8$

$\frac{0.4}{\quad}x \leq \frac{0.8}{\quad}$ Divide both sides by .

$x \leq$

Example 3

PROBLEM SOLVING APPLICATION

With first-class mail, there is an extra charge in any of these cases:

- The length is greater than $11\frac{1}{2}$ inches.
- The height is greater than $6\frac{1}{8}$ inches.
- The thickness is greater than $\frac{1}{4}$ inch.
- The length divided by the height is less than 1.3 or greater than 2.5.

The height of an envelope is 3.8 in. What are the minimum and maximum lengths to avoid an extra charge?

1. Understand the Problem

The answer is the ______ and ______ lengths for an envelope to avoid an extra charge. List the important information:

- The height of the piece of mail is ______ inches.
- If the length divided by the height is between ______ and ______, there will not be an extra charge.

Show the relationship of the information:

$\square \le \dfrac{\text{length}}{\text{height}} \le \square$

2. Make a Plan

You can use the model of the relationship to write an inequality where l is the length and ______ is the height.

$\square \le \dfrac{l}{3.8} \le \square$

3. Solve

$1.3 \le \dfrac{l}{3.8}$ and $\dfrac{l}{3.8} \le 2.5$

$\square \cdot 1.3 \le l$ and $l \le 2.5 \cdot \square$ — Multiply both sides of each inequality by ______.

$l \ge$ ______ and $l \le$ ______ — Simplify.

The length of the envelope must be between ______ in. and ______ in.

4. Look Back

$4.94 \div 3.8 =$ ______ and $9.5 \div 3.8 =$ ______, so there will not be an extra charge.

Try This

1. Solve.

$0.6y \leq 1.8$

2. Problem Solving Application

With first-class mail, there is an extra charge in any of these cases:

- The height is greater than $6\frac{1}{2}$ inches.
- The length is greater than $11\frac{1}{8}$ inches.
- The thickness is greater than $\frac{1}{4}$ inch.
- The length divided by the height is less than 1.3 or greater than 2.5.

The height of an envelope is 4.9 in. What are the minimum and maximum lengths to avoid an extra charge?

LESSON 3-8

Squares and Square Roots

pp. 146–147

Vocabulary

principal square root (p. 146) ______

perfect square (p. 146) ______

Additional Examples

Example 1

Find the two square roots of each number.

A. 49

$\sqrt{49} =$ ___ ___ is a square root, since $7 \cdot 7 =$ ___.

$-\sqrt{49} =$ ___ ___ is also a square root, since $-7 \cdot (-7) =$ ___.

B. 100

$\sqrt{100} =$ ___ ___ is a square root, since $10 \cdot 10 =$ ___.

$-\sqrt{100} =$ ___ ___ is also a square root, since $-10 \cdot (-10) =$ ___.

C. 225

$\sqrt{225} =$ ___ ___ is a square root, since $15 \cdot 15 =$ ___.

$-\sqrt{225} =$ ___ ___ is also a square root, since $-15 \cdot (-15) =$ ___.

Example 3

Evaluate each expression.

A. $3\sqrt{36} + 7$

$3\sqrt{36} + 7 = 3(\quad) + 7$ Evaluate the root.

$= \quad + 7$ Multiply.

$=$ Add.

B. $\sqrt{21 - 5} + 9$

$\sqrt{21 - 5} + 9 = \sqrt{\quad} + 9$ Evaluate the expression under the square symbol.

$= \quad + 9$ Evaluate the root.

$=$ Add.

Try This

1. Find the two square roots of the number.

144

2. Evaluate the expression.

$2\sqrt{25} + 4$

LESSON 3-9
Finding Square Roots
pp. 150–151

Additional Examples

Example 2

PROBLEM SOLVING APPLICATION

You want to sew a fringe on a square tablecloth with an area of 500 square inches. Calculate the length of each side of the tablecloth and the length of fringe you will need to the nearest tenth of an inch.

1. Understand the Problem

First find the length of a side. Then you can use the length of a side to find the ______, the length of fringe around the tablecloth.

2. Make a Plan

The length of a side, in inches, is the number that you multiply by itself to get ______. To be accurate, find this number to the nearest tenth. If you do not know a step-by-step method for finding $\sqrt{500}$, use guess and check.

3. Solve

Because 500 is between 22^2 and 23^2, the square root of 500 is between ______ and ______.

Guess 22.5	Guess 22.2	Guess 22.4	Guess 22.3
$22.5^2 = 506.25$	$22.2^2 = 492.84$	$22.4^2 = 501.76$	$22.3^2 = 497.29$
Too high	Too low	Too high	Too low
Square root is between 22 and 22.5	Square root is between 22.2 and 22.5	Square root is between 22.2 and 22.4	Square root is between 22.3 and 22.4

The square root is between 22.3 and 22.4. To round to the nearest tenth, look at the next decimal place. Consider 22.35.

$22.35^2 =$ ______ Too ______

The square root must be greater than 22.35, so round up.

To the nearest tenth, $\sqrt{500}$ is about ______.

The length of each side of the tablecloth is about in.

The length of a side of the tablecloth is inches, to the nearest tenth of an inch. Now estimate the length around the tablecloth.

$\cdot$ 22.4 = Perimeter = 4 $\cdot$ side

You will need about inches of fringe.

4. Look Back

The length 90 inches divided by 4 is inches. A 22.5-inch square has an area of square inches, which is close to 500, so the answers are reasonable.

Try This

The square root is between two integers. Name the integers.

1. $\sqrt{80}$

2. $-\sqrt{45}$

LESSON 3-10

The Real Numbers

pp. 156–157

Know-It Notes

Vocabulary

irrational number (p. 156)

real number (p. 156)

Density Property (p. 157)

Additional Examples

Example 1

Write all names that apply to each number.

A. $\sqrt{5}$ 5 is a ______ number that is not a perfect ______.

B. -12.75 -12.75 is a ______ decimal.

Example 2

State if the number is rational, irrational, or not a real number.

A. $\sqrt{15}$ 15 is a ______ number that is not a perfect ______.

B. $\frac{0}{3}$ $\frac{0}{3} =$ ______

Example 3

Find a real number between $3\frac{2}{5}$ and $3\frac{3}{5}$.

There are many solutions. One solution is halfway between the two numbers. To find it, add the numbers and divide by 2.

$\left(3\frac{2}{5} + 3\frac{3}{5}\right) \div 2 = \quad \frac{\quad}{5} \div 2 = \quad \div 2 = \quad .$

3   $3\frac{1}{5}$ $3\frac{2}{5}$ $3\frac{3}{5}$ $3\frac{4}{5}$ 4

A real number between $3\frac{2}{5}$ and $3\frac{3}{5}$ is   .

Try This

1. Write all names that apply to the number.

$\sqrt{9}$

2. State if the number is rational, irrational, or not a real number.

$\sqrt{-7}$

3. Find a real number between $4\frac{3}{7}$ and $4\frac{4}{7}$.

Chapter 3
Vocabulary Chain

EXAMPLE		TERM
		irrational number

EXAMPLE		TERM
		perfect square

EXAMPLE		TERM
		rational number

EXAMPLE		TERM
		real number

EXAMPLE		TERM
		relatively prime

Directions

1. Write an example and an explanation in words, numbers, and algebra for each term.
2. Cut out each vocabulary strip and fold in thirds.
3. Punch a hole in the corner of each folded vocabulary strip, and string them together to create your vocabulary chain.

Chapter 3
Vocabulary Chain

WORDS	NUMBERS	ALGEBRA
WORDS	NUMBERS	ALGEBRA
WORDS	NUMBERS	ALGEBRA
WORDS	NUMBERS	ALGEBRA
WORDS	NUMBERS	ALGEBRA

LESSON 4-1

Samples and Surveys
pp. 174–175

Vocabulary

population (p. 174)

sample (p. 174)

biased sample (p. 174)

random sample (p. 175)

systematic sample (p. 175)

stratified sample (p. 175)

Additional Examples

Example 1

Identify the population and the sample. Give a reason why the sample could be biased.

A. A record store manager asks customers who make a purchase how many hours of music they listen to each day.

Population	Sample	Possible Bias
		Customers who make a purchase might be more ____ in music than others in the store.

Example 2

Identify the sampling method used.

A. In a county survey, Democratic Party members whose names begin with the letter D are chosen.

The ______ is to survey members whose names begin with D.

B. A telephone company randomly chooses customers to survey about its service.

Customers are chosen by ______.

Try This

1. Identify the population and the sample. Give a reason why the sample could be biased.

People attending a baseball game were asked if they support the construction of a new stadium in the city.

Population	Sample	Possible Bias
		People that attend a baseball game are more likely to the construction of a new stadium.

2. Identify the sampling method used.

In a county survey, families with 3 or more children are chosen.

LESSON 4-2

Organizing Data

pp. 179–180

Know-It Notes

Vocabulary

stem-and-leaf plot (p. 179)

back-to-back stem-and-leaf plot (p. 180)

Additional Examples

Example 1

Use the given data to make a table.

Jack timed his bus rides to and from school. On Monday, it took 7 minutes to get to school and 9 minutes to get home. On Tuesday, it took 5 minutes and 9 minutes, respectively, and on Wednesday, it took 8 minutes and 7 minutes.

Example 2

List the data values of the stem-and-leaf plot.

Stem	Leaves
1	2 5
4	0 1 1
5	2 7 9

Key: 1|2 means 12

The data values are

Example 3

Use the given data to make a stem-and-leaf plot.

Top Speeds of Animals (mi/h)			
Cheetah	64	Elk	45
Wildebeest	61	Coyote	43
Lion	50	Gray Fox	42

Speeds range from 42 to 64 so stems are 4 to 6.

Example 4

Use the given data to make a back-to-back stem-and-leaf plot.

U.S. Representatives for Selected States, 1950 and 2000

	IL	MA	MI	NY	PA
1950	25	14	18	43	31
2000	19	10	15	29	19

Try This

1. Use the given data to make a table.

Jill timed herself jogging to the park and back home. On Monday she ran to the park in 12 minutes then back home in 14 minutes, on Tuesday it took her 13 and 15 minutes, respectively, and on Wednesday, it took her 11 minutes and 13 minutes.

2. List the data values of the stem-and-leaf plot.

Stem	Leaves
2	3 6
3	7 8 9
4	2 5 6

Key: 2|3 means 23

3. Use the given data to make a stem-and-leaf plot.

Percent of Persons under 18 years old, year 2000					
Florida	California	Texas	Arizona	New York	Alaska
23%	27%	28%	27%	25%	30%

4. Use the given data to make a back-to-back stem-and-leaf plot.

U.S. Representatives for Selected States, 1950 and 2000					
	IL	MA	MI	NY	PA
1950	25	14	18	43	31
2000	19	10	15	29	19

LESSON 4-3

Measures of Central Tendency
pp. 184–185

Know-It Notes

Vocabulary

mean (p. 184)

median (p. 184)

mode (p. 184)

outlier (p. 185)

Additional Examples

Example 1

Find the mean, median, and mode of the data set.

A. 16, 25, 31, 14, 14, 18

mean: 16 + 25 + 31 + 14 + 14 + 18 = the values.

$\frac{118}{} \approx$ Divide by , the number of values.

median: 14 14 16 | 18 25 31 the values.

3 values *3 values*

$\frac{ + }{2} =$ the two middle values.

mode: The value occurs two times.

B. 83, 45, 19, 33

mean: $83 + 45 + 19 + 33 = \square$ $\square$ the values.

$\frac{180}{\square} =$ Divide by $\square$, the number of values.

median: 19 33 | 45 83 $\square$ the values.

2 values *2 values*

$\frac{\square + \square}{2} = \square$ Average the two $\square$ values.

mode: $\square$ $\square$ occurs more than any other.

Try This

1. Find the mean, median, and mode of the data set.

24, 31, 21, 18, 24, 22

LESSON **4-4**

Variability

pp. 188–190

Vocabulary

variability (p. 188)

range (p. 188)

quartile (p. 188)

box-and-whisker plot (p. 189)

Additional Examples

Example 2

Use the given data to make a box-and-whisker plot.

21, 25, 15, 13, 17, 19, 19, 21

Step 1. Order the data and find the smallest value, first quartile, median, third quartile, and largest value.

smallest value:

first quartile: $\dfrac{\quad + \quad}{2} =$

median: $\dfrac{\quad + \quad}{2} =$

third quartile: $\dfrac{\quad + \quad}{2} =$

largest value:

Step 2. Draw a number line and plot a point above each value from Step 1.

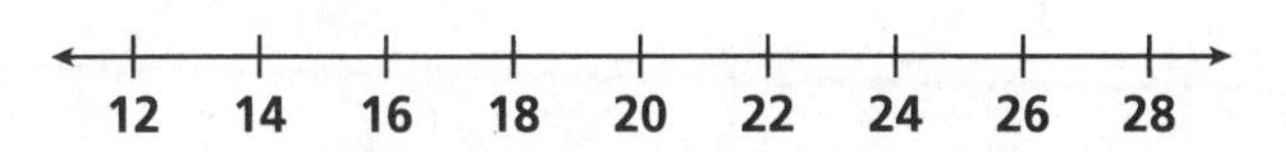

Step 3. Draw the box and whiskers.

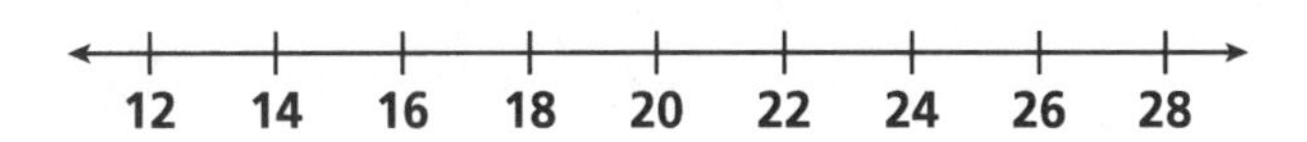

Try This

1. Find the range and the first and third quartiles for the data set.

45, 31, 59, 49, 49, 69, 33, 47

2. Use the given data to make a box-and-whisker plot.

31, 23, 33, 35, 26, 24, 31, 29

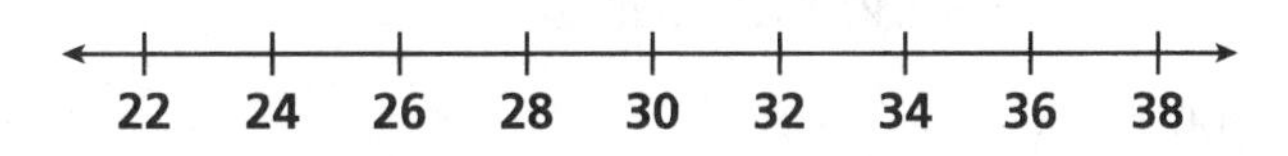

LESSON 4-5

Displaying Data

pp. 196–197

Vocabulary

bar graph (p. 196)

frequency table (p. 196)

histogram (p. 196)

line graph (p. 197)

Additional Examples

Example 1

Organize the data into a frequency table and make a bar graph.

The following data set reflects the number of hours of television watched every day by members of a sixth-grade class:

1 1 3 0 2 0 5 3 1 3

First, organize the data into a table.

The is the number of times each value occurs.

The frequencies are the of the bars in the bar graph.

Example 2

Jimmy surveyed 12 children to find out how much money they received from the tooth fairy. Use the data to make a histogram.

0.35 2.00 0.75 2.50 1.50 3.00 0.25 1.00 1.00 3.50 0.50 3.00

First, make a ______ table with intervals of $1.00.

Then make a ______.

Example 3

Make a line graph of the given data. Use the graph to estimate Mr. Yi's salary in 1992.

Year	Salary ($)
1985	42,000
1990	49,000
1995	58,000
2000	69,000

Create ordered pairs from the data in the table and plot them on a grid. Connect the points with lines. You can estimate the salary in 1992 by finding the point on the line between 1990 and 1995 that corresponds to 1992.

Mr. Yi's 1992 salary was about $______.

Try This

1. Organize the data into a frequency table and make a bar graph.

The following data set reflects the number of laptop computers that are repaired by Mike the technician in one week. (Each number reflects one day.)

3 5 7 8 4

2. Tonya surveyed 14 children at an after school day care to find out how many hours they spend there. Use the data to make a histogram.

2:00 1:40 3:00 0:30 1:00 1:30 2:30 0:30 2:45 1:00 2:00 1:35 1:30 3:00

LESSON **4-6**

Misleading Graphs and Statistics

pp. 200–201

Additional Examples

Example 1

Explain why each graph is misleading.

A.

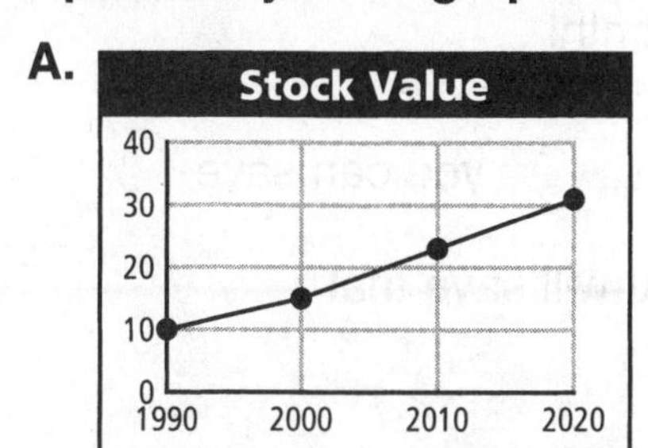

The graph suggests that the stock will continue to increase through _____, but there's no way to foresee the _____.

B.

School Play
Ticket Sales
Tickets Sold
120
110
100
0
6th
7th
8th
Grade

Because the first interval of the scale goes from _____ to _____, the bar _____ make it appear that the sixth grade sold about _____ times as many tickets as either of the other two grades. In fact, the sixth grade sold only about _____ more.

Example 2

Explain why each statistic is misleading.

A. Sam scored 43 goals for his soccer team during the season, and Jacob scored only 2.

Although Jacob scored only _____ goals, he may have played most of his time on _____.

B. Four out of five dentists surveyed preferred UltraClean toothpaste.

This statement does not give the ______ size or state what UltraClean toothpaste was ______ with.

C. Shopping at Save-a-Lot can save you up to $100 a month!

The words *save up to $100* mean that the ______ you can save is $100, but there is no ______ that you will save that amount.

Try This

1. Explain why the graph is misleading.

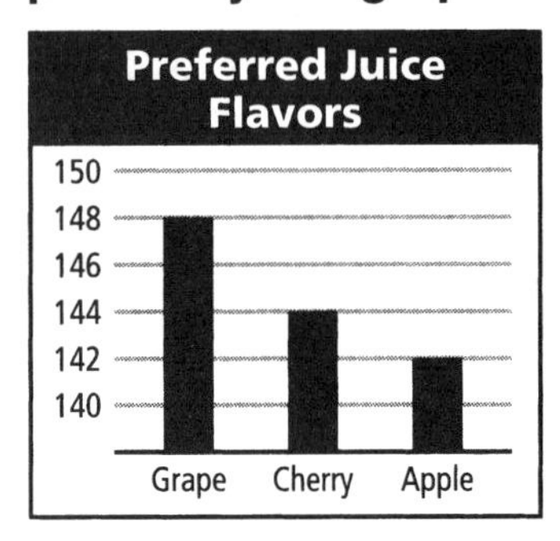

2. Explain why the statistic is misleading.

The total revenue for bathing suits sold in May at Worthman's Florida stores is $250,000. The total revenue for bathing suits sold in May at Worthman's North Dakota stores is $10,000.

LESSON **4-7**

Scatter Plots

pp. 204–205

Vocabulary

scatter plot (p. 204)

correlation (p. 204)

line of best fit (p. 204)

Additional Examples

Example 1

Use the given data to make a scatter plot of the weight and height of each member of a basketball team.

Height (in.)	Weight (lb)
71	170
68	160
70	175
73	180
74	190

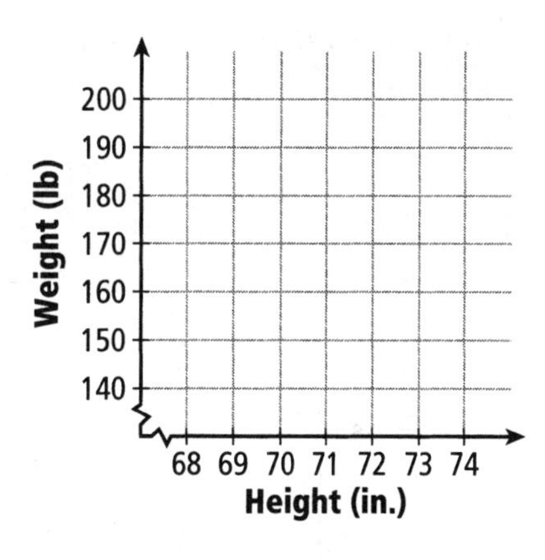

The points on the scatter plot are ( ), ( ),

( ), ( ), and ( ).

Example 2

Do the data sets have a positive, a negative, or no correlation?

A. The size of a jar of baby food and the number of jars of baby food a baby will eat.

correlation: The food in each jar, the

number of jars of baby food a baby will eat.

B. The speed of a runner and the number of races she wins.

correlation: The the runner, the

races she will win.

Try This

1. Use the given data to make a scatter plot of the weight and height of each member of a soccer team.

Height (in)	Weight (lb)
63	125
67	156
69	175
68	135
62	120

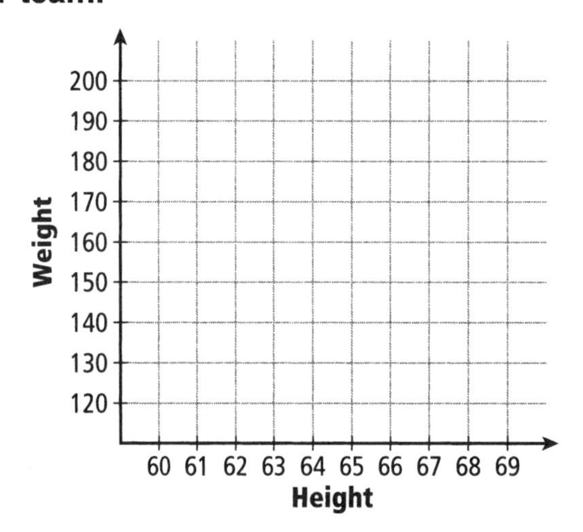

2. Does the data set have a positive, a negative, or no correlation?

Your grade point average and the number of A's you receive.

Chapter 4
Picture Cube

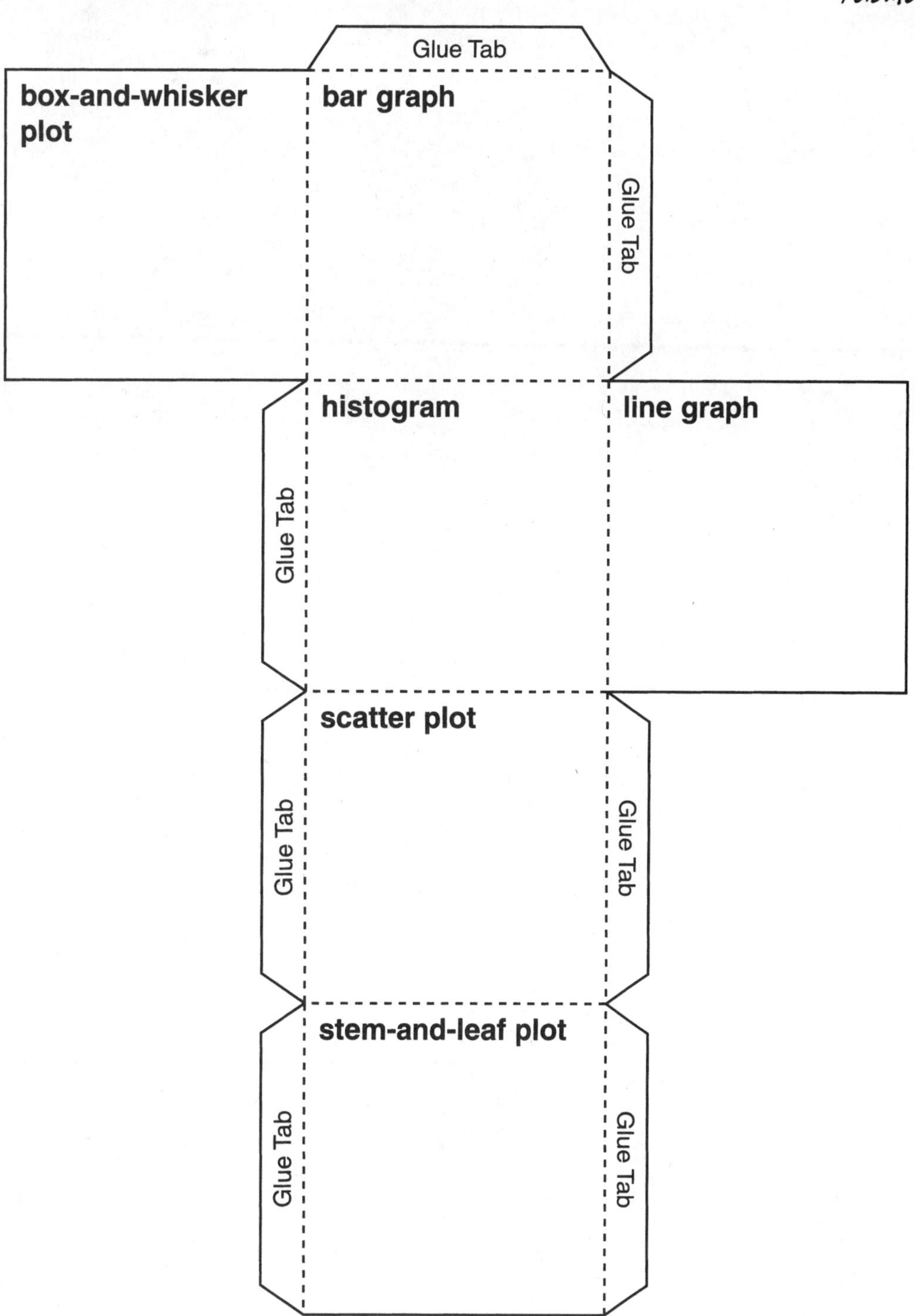

Directions

1. Draw an example of each term on the net of the cube.
2. Cut out the net.
3. Fold along all dotted lines, and place glue tabs to the inside of the cube.
4. Join the common edges, and tape or glue the tabs in place.

LESSON **5-1**

Points, Lines, Planes, and Angles

pp. 222–224

Vocabulary

point (p. 222)

line (p. 222)

plane (p. 222)

segment (p. 222)

ray (p. 222)

angle (p. 222)

right angle (p. 223)

acute angle (p. 223)

obtuse angle (p. 223)

complementary angles (p. 223)

supplementary angles (p. 223)

vertical angles (p. 223)

congruent (p. 223)

Additional Examples

Example 1

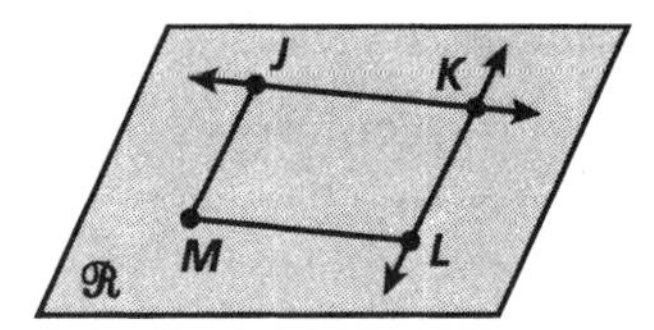

A. Name 4 points in the figure.

B. Name a line in the figure.

Any points on a line can be used.

C. Name a plane in the figure.

Any points in the plane that form a triangle can be used.

Example 2

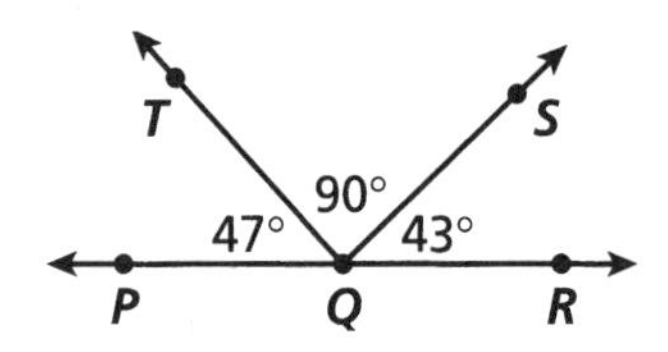

A. Name a right angle in the figure.

B. Name two acute angles in the figure.

Example 3

In the figure, $\angle 1$ and $\angle 3$ are vertical angles, and $\angle 2$ and $\angle 4$ are vertical angles.

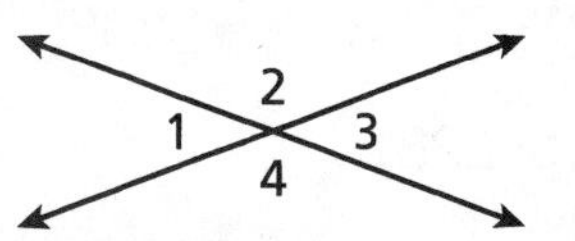

A. If $m\angle 1 = 37°$, find $m\angle 3$.

The measures of $\angle 1$ and $\angle 2$ add to 180° because they are ______, so $m\angle 2$ = ____° − ____° = ____°.

The measures of $\angle 2$ and $\angle 3$ add to 180° because they are ______, so $m\angle 3$ = ____° − ____° = ____°.

B. If $m\angle 4 = y°$, find $m\angle 2$.

$m\angle 3 = 180° - y°$

$m\angle 2 = 180° - (180° - y°)$

$= 180° - 180° + y°$ Distributive Property

$=$ ____ $m\angle 2$ ____ $m\angle 4$

Try This

1.

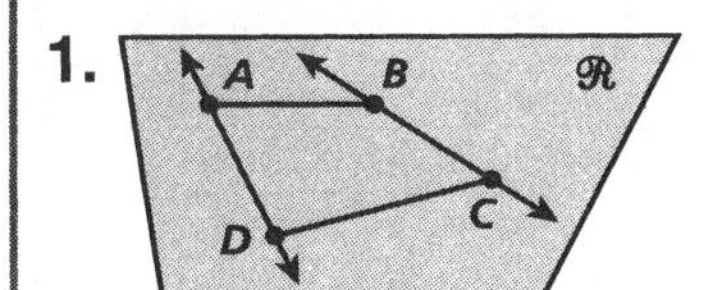

Name a line in the figure.

______

2.

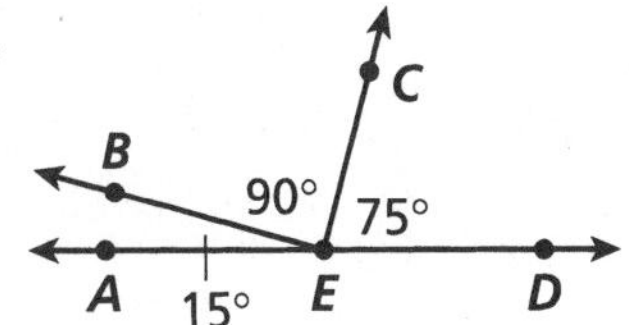

Name a pair of complementary angles.

______

LESSON 5-2

Parallel and Perpendicular Lines

pp. 228–229

Vocabulary

parallel lines (p. 228)

perpendicular lines (p. 228)

transversal (p. 228)

Additional Examples

Example 1

Measure the angles formed by the transversal and parallel lines. Which angles seem to be congruent?

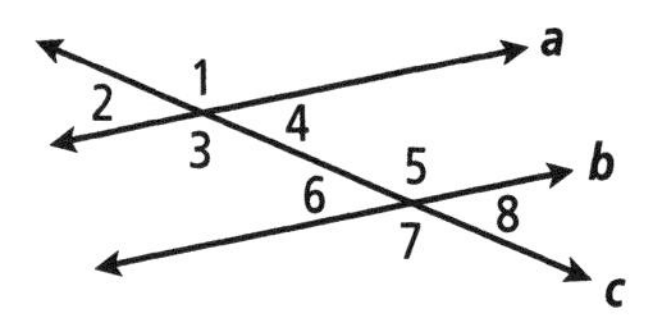

∠1, ∠3, ∠5, and ∠7 all measure 150°.

∠2, ∠4, ∠6, and ∠8 all measure °.

Example 2

In the figure, line *l* ∥ line *m*.
Find the measure of each angle.

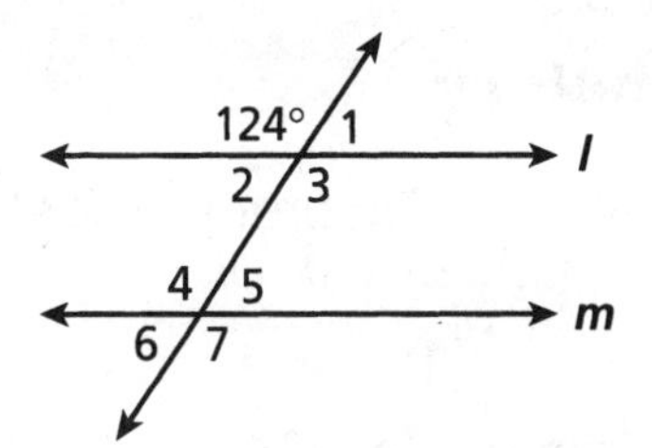

A. ∠4

m∠4 = ____°

All obtuse angles in the figure are ____.

B. ∠2

m∠2 + 124° = ____°

∠2 is ____ to the angle 124°.

− ____°   − ____°

m∠2 = ____°

C. ∠6

m∠6 = ____°

All acute angles in the figure are ____.

Try This

1. Measure the angles formed by the transversal and parallel lines. Which angles seem to be congruent?

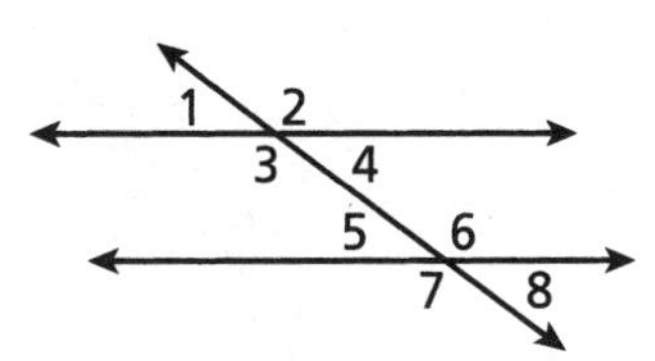

2. In the figure, line *n* ∥ line *m*. Find the measure of the angle. ∠7

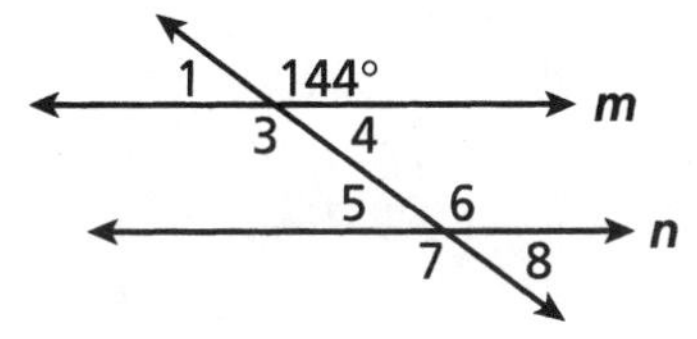

LESSON 5-3

Triangles

pp. 234–236

Vocabulary

Triangle Sum Theorem (p. 234)

acute triangle (p. 234)

right triangle (p. 234)

obtuse triangle (p. 234)

equilateral triangle (p. 235)

isosceles triangle (p. 235)

scalene triangle (p. 235)

Additional Examples

Example 1

A. Find p in the acute triangle.

$\_\_\_^\circ + \_\_\_^\circ + p = \_\_\_^\circ$

$\_\_\_^\circ + p = \_\_\_^\circ$

$- 117^\circ \qquad - 117^\circ$

$p = \_\_\_^\circ$

p°, 73°, 44°

B. Find c in the right triangle.

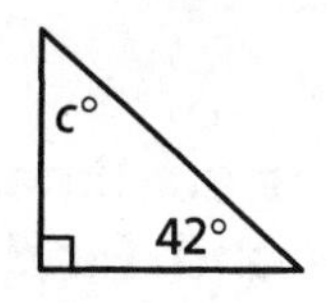

$\_\_\_^\circ + \_\_\_^\circ + c = \_\_\_^\circ$

$\_\_\_^\circ + c = \_\_\_^\circ$

$-132^\circ \qquad -132^\circ$

$c = \_\_\_^\circ$

Example 2

A. Find the angle measures in the equilateral triangle.

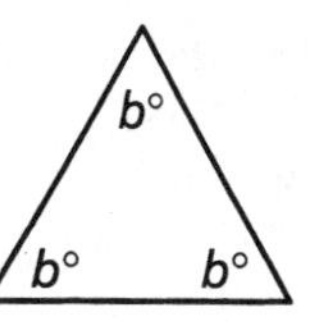

$3b^\circ = \_\_\_^\circ$ Triangle ___ Theorem

$\frac{3b^\circ}{\_\_\_} = \frac{180^\circ}{\_\_\_}$ Divide both sides by ___.

$b^\circ = \_\_\_^\circ$

B. Find the angle measures in in the isosceles triangle.

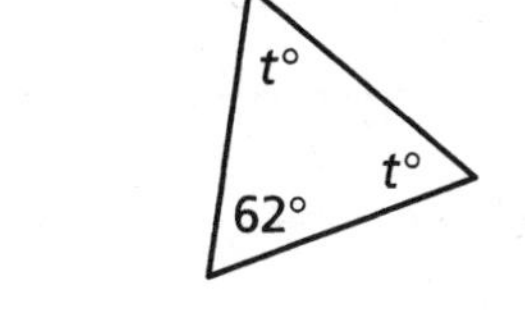

$\_\_\_^\circ + t^\circ + t^\circ = \_\_\_^\circ$ ___ Sum Theorem

$62^\circ + 2t^\circ = \_\_\_^\circ$ Combine like terms.

$-62^\circ \qquad -62^\circ$ Subtract 62° from both sides.

$2t^\circ = \_\_\_^\circ$

$\frac{2t^\circ}{\_\_\_} = \frac{118^\circ}{\_\_\_}$ Divide both sides by ___.

$t^\circ = \_\_\_^\circ$

Example 3

The second angle in a triangle is six times as large as the first. The third angle is half as large as the second. Find the angle measures and draw a possible picture.

Let $x°$ = the first angle measure. Then $6x°$ = second angle measure, and $\frac{1}{2}(6x°) = 3x°$ = third angle measure.

$x° + 6x° + 3x° =$ ° Triangle Theorem

$\frac{10x°}{} = \frac{180°}{}$ Combine terms. Divide both sides by .

$x° =$ °

The angles measure °, °, and °. The triangle is an obtuse triangle.

Try This

1. Find *b* in the right triangle.

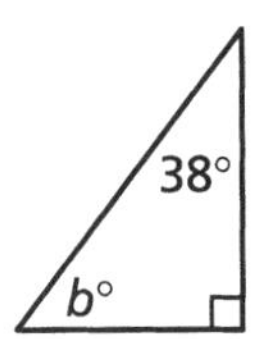

2. Find the angle measures in the isosceles triangle.

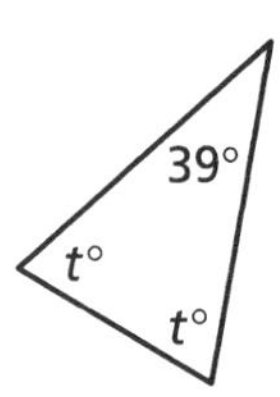

LESSON **5-4**

Polygons

pp. 239–241

Vocabulary

polygon (p. 239)

regular polygon (p. 240)

trapezoid (p. 240)

parallelogram (p. 240)

rectangle (p. 240)

rhombus (p. 240)

square (p. 240)

Additional Examples

Example 1

Find the sum of the angle measures in the figure.

Find the sum of the angle measures in a hexagon.

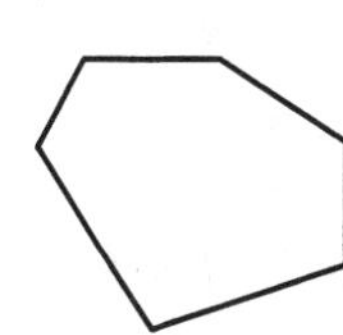

Divide the figure into ____________ .

____ · 180° = ____° ____ triangles

Example 2

Find the angle measures in the regular polygon.

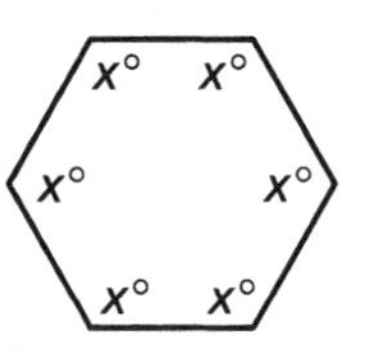

6 congruent angles

$6x = 180°(6 - 2)$

$6x = 180°(\quad)$

°

$6x =$

$\frac{6x}{\quad} = \frac{720°}{\quad}$

°

$x =$

Example 3

Give all the names that apply to each figure.

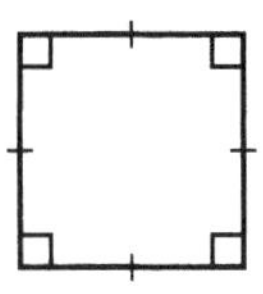

-sided polygon

2 pairs of sides

4 angles

4 sides

4 sides and

4 angles

Try This

1. Find the sum of the angle measures in the figure.

Find the sum of the angle measures in a heptagon.

2. Find the angle measures in the regular polygon.

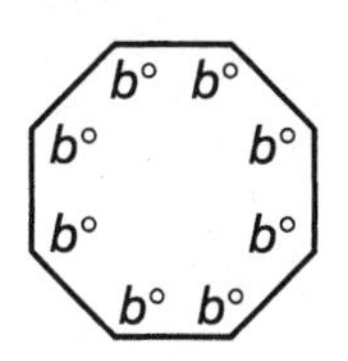

3. Give all the names that apply to the figure.

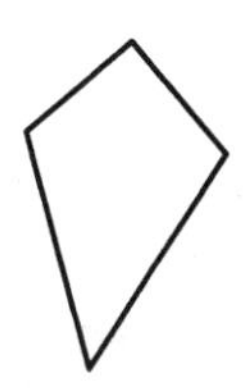

LESSON 5-5 Coordinate Geometry

pp. 244–246

Know-It Notes

Vocabulary

slope (p. 244)

rise (p. 244)

run (p. 244)

Additional Examples

Example 1

Determine if the slope of each line is positive, negative, 0, or undefined. Then find the slope of each line.

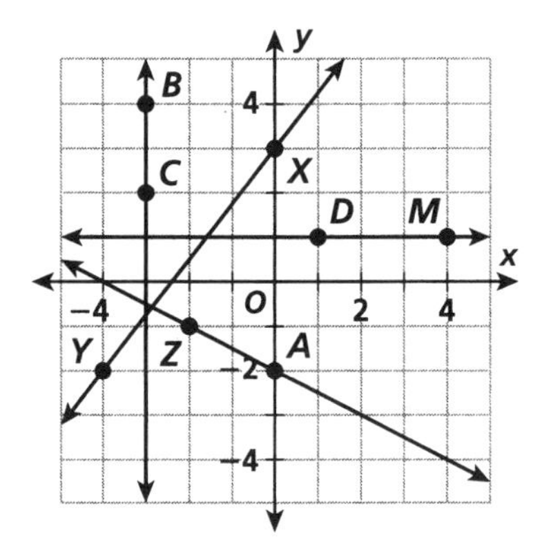

A.  $\overleftrightarrow{XY}$

slope;

slope of $\overleftrightarrow{XY}$ = ——— =

B.  $\overleftrightarrow{ZA}$

slope;

slope of $\overleftrightarrow{ZA}$ = ——— =

C. $\overleftrightarrow{BC}$

slope of $\overleftrightarrow{BC}$ is

Example 2

Which lines are parallel?
Which lines are perpendicular?

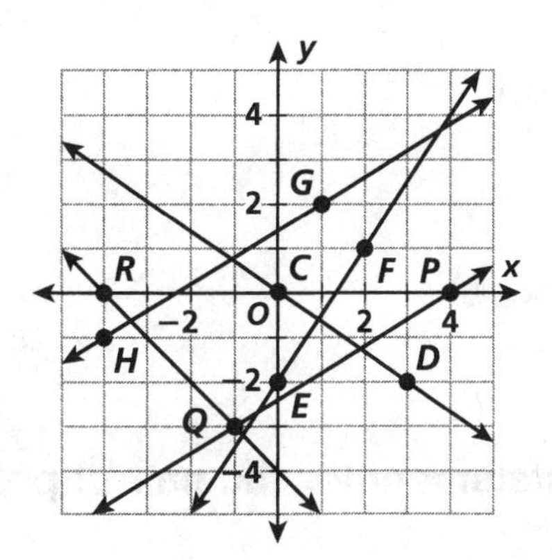

slope of $\overleftrightarrow{EF} = \frac{3}{\square}$

slope of $\overleftrightarrow{GH} = \frac{3}{\square}$

slope of $\overleftrightarrow{PQ} = \frac{3}{\square}$

slope of $\overleftrightarrow{CD} = \frac{\square}{\square}$ or $\square$

slope of $\overleftrightarrow{QR} = \frac{\square}{\square}$ or $\square$

$\square \parallel \square$ The slopes are $\square$. $\frac{3}{5} = \frac{3}{5}$

$\square \perp \square$ The slopes have a $\square$ of −1: $\frac{3}{2} \cdot -\frac{2}{3} = -1$

Try This

1. Which lines are parallel? Which lines are perpendicular?

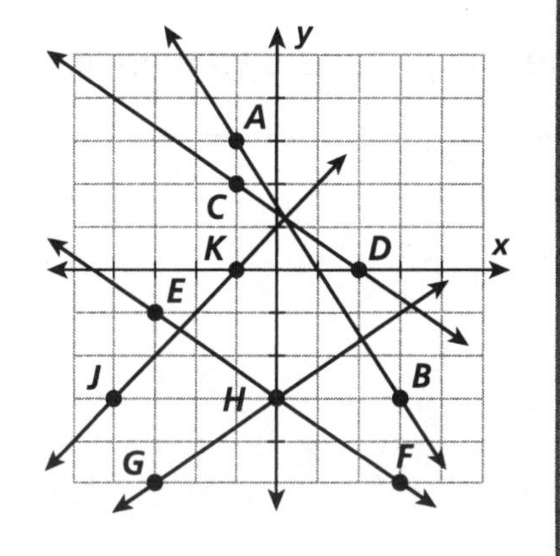

LESSON **5-6**

Congruence

pp. 250–251

Know-It Notes

Vocabulary

correspondence (p. 250)

Additional Examples

Example 1

Write a congruence statement for the pair of polygons.

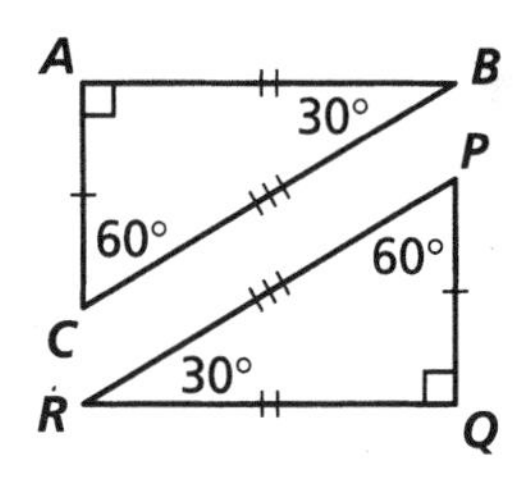

The first triangle can be named triangle *ABC*. To complete the congruence statement, the vertices in the second triangle have to be written in

of the .

$\angle A \cong \angle$, so $\angle A$ corresponds to $\angle$.

$\angle B \cong \angle$, so $\angle B$ corresponds to $\angle$.

$\angle C \cong \angle$, so $\angle C$ corresponds to $\angle$.

The congruence statement is triangle $\cong$ triangle .

Example 2

In the figure, quadrilateral *VWXY* ≅ quadrilateral *JKLM*.

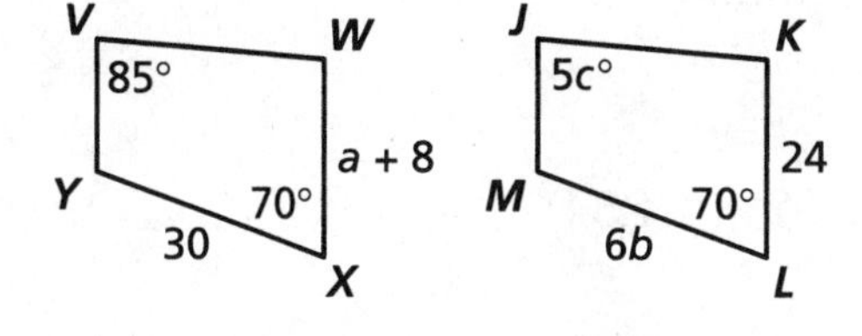

A. Find *a*.

$a + 8 = 24$ $\quad \overline{WX} \cong$ ______

$\underline{-8} \quad \underline{-8}$ $\quad$ Subtract 8 from both sides.

$a =$ ______

B. Find *b*.

$6b = 30$ $\quad \overline{ML} \cong$ ______

$\frac{6b}{6} = \frac{30}{6}$ $\quad$ Divide both sides by 6.

$b =$ ______

Try This

1. Write a congruence statement for the pair of polygons.

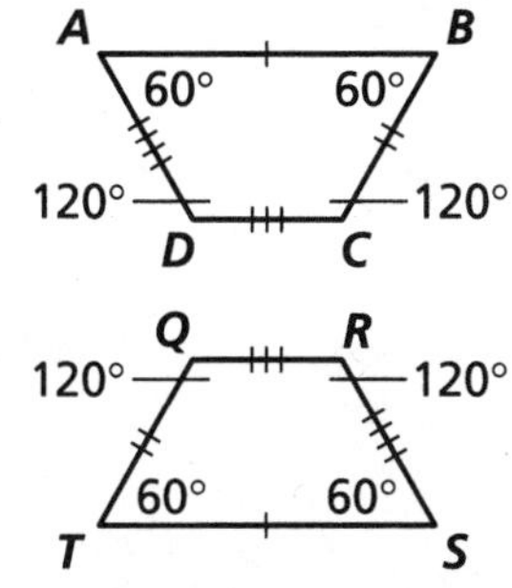

2. In the figure, quadrilateral *JIHK* ≅ quadrilateral *QRST*.
Find *c*.

H 3a I
4b°
90°
J
30°
K
R 6 S
90°
120°
Q
c + 10°
T

LESSON 5-7 **Transformations**
pp. 254–255

Vocabulary

transformation (p. 254)

translation (p. 254)

rotation (p. 254)

center of rotation (p. 254)

reflection (p. 254)

image (p. 254)

Additional Examples

Example 1

Identify as a translation, rotation, reflection, or none of these.

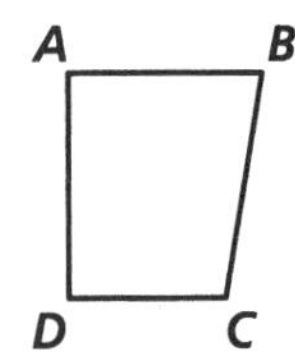

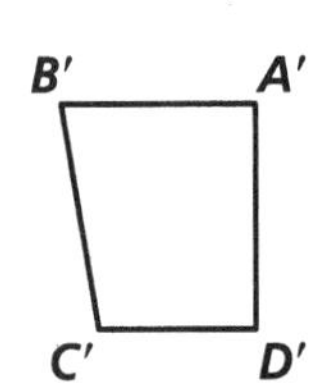

Example 2

Draw the image of the triangle after the transformation.

Translation along $\overline{AB}$ so that A' coincides with B

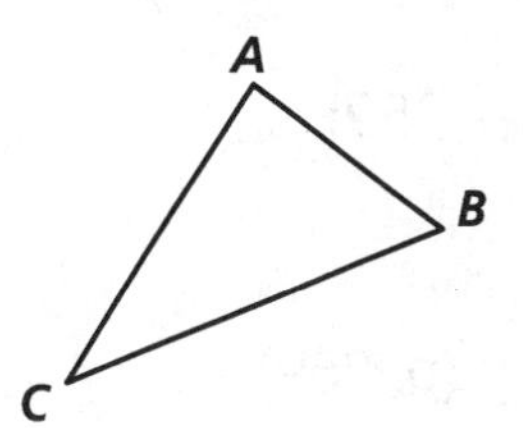

Example 3

Draw the image of a triangle with vertices of (1, 1), (2, −2), and (5, 0) after the transformation.

180° counterclockwise rotation around (0, 0)

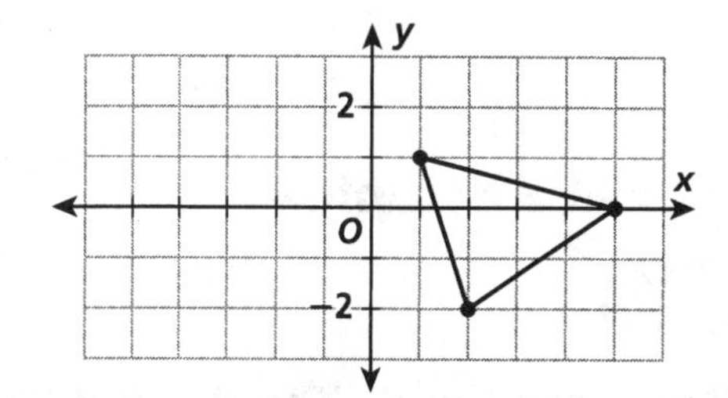

Try This

1. Identify the transformation as a translation, rotation, reflection, or none of these.

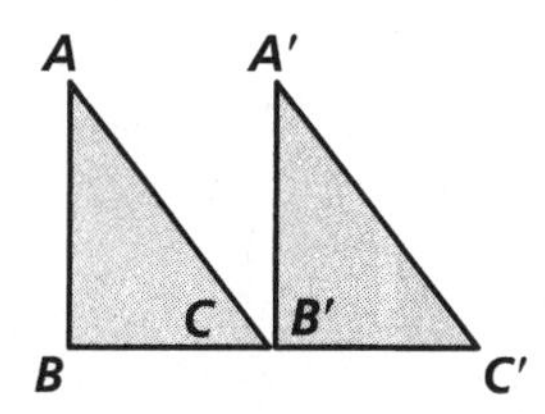

2. Draw the image of the polygon after the transformation.

90° counterclockwise rotation around point C

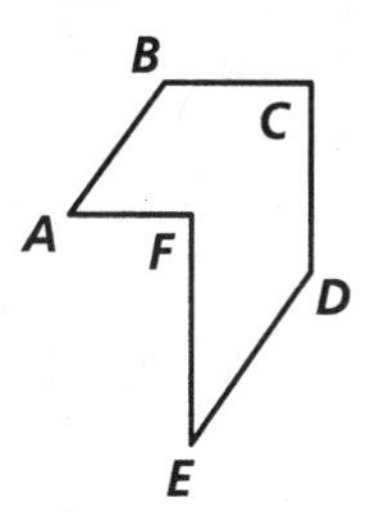

LESSON **5-8**

Symmetry
pp. 259–260

Vocabulary

line symmetry (p. 259)

line of symmetry (p. 259)

rotational symmetry (p. 260)

Additional Examples

Example 1

Complete each figure. The dashed line is the line of symmetry.

A.

B.

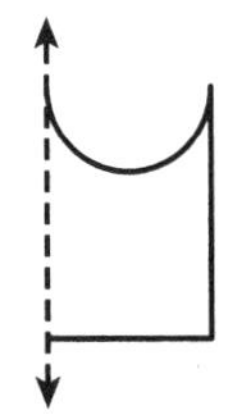

Example 2

Complete each figure. The point is the center of rotation.

A. 2-fold

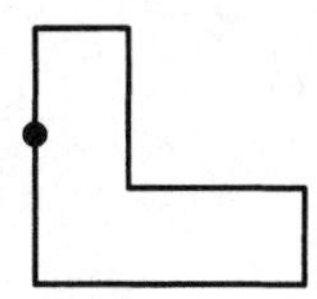

The figure coincides with itself every ______.

B. 5-fold

The figure coincides with itself every ______.

Try This

1. Complete the figure. The dashed line is the line of symmetry.

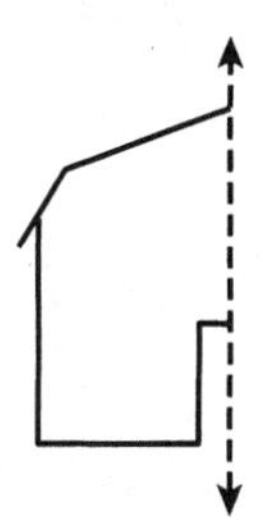

2. Complete the figure. The point is the center of rotation.

8-fold

LESSON **5-9**

Tessellations

pp. 263–265

Vocabulary

tessellation (p. 263)

regular tessellation (p. 263)

semiregular tessellation (p. 263)

Additional Examples

Example 1

PROBLEM SOLVING APPLICATION

Find all the possible semiregular tessellations that use triangles and squares.

1. **Understand the Problem**
   List the important information:
   - The angles at each vertex add to 360°.
   - All the angles in a square measure °.
   - All the angles in an equilateral triangle measure °.
2. **Make a Plan**
   Account for all possibilities: List all possible combinations of triangles and squares around a vertex that add to 360°. Then see which combinations can be used to create a semiregular tessellation.

6 triangles, 0 squares	6(60°) = °	regular
3 triangles, 2 squares	3(60°) + 2(90°) = °	
0 triangles, 4 squares	4(90°) = °	regular

3. Solve

There are two arrangements of three triangles and two squares around a vertex.

Repeat each arrangement around every vertex, if possible, to create a tessellation.

There are exactly two semiregular tessellations that use triangles and squares.

4. Look Back

Every vertex in each arrangement is identical to the other vertices in that arrangement, so these are the only arrangements that produce semiregular tessellations.

Example 2

Create a tessellation with quadrilateral *EFGH*.

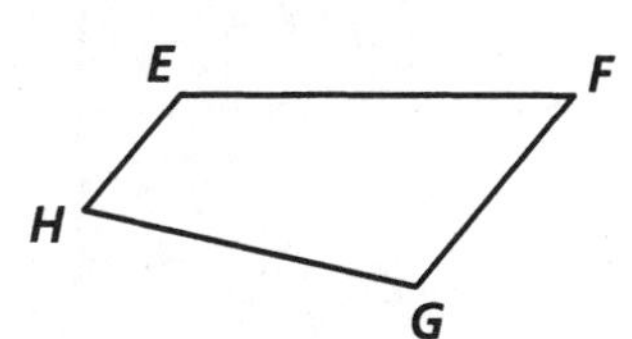

There must be a copy of each angle of quadrilateral *EFGH* at every vertex.

Example 3

Use rotations to create a tessellation with the quadrilateral given below.

Step 1: Find the midpoint of a side.

Step 2: Make a new edge for half of the side.

Step 3: Rotate the new edge around the midpoint to form the edge of the other half of the side.

Step 4: Repeat with the other sides.

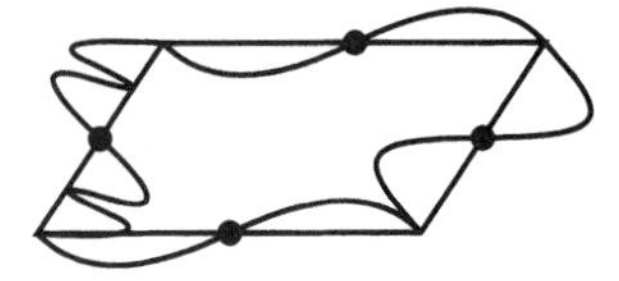

Step 5: Use the figure to make a tessellation.

Try This

1. Create a tessellation with quadrilateral *IJKL*.

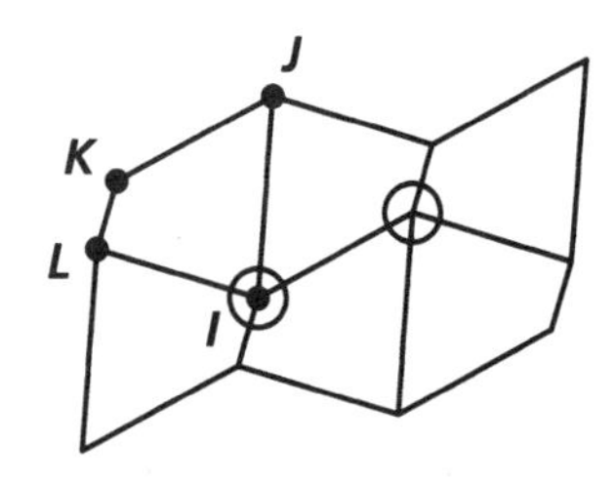

2. Create a tessellation with the figure given below.

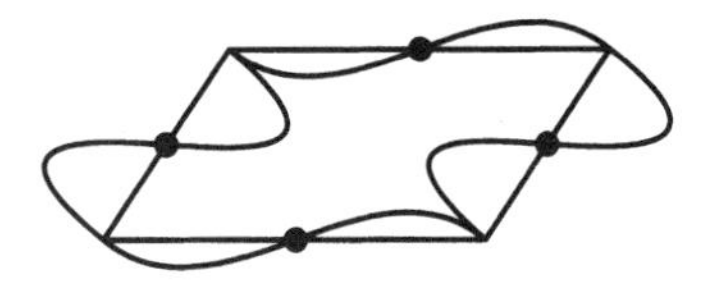

Chapter 5
Möbius Mobile

acute triangle ______________________________
complementary angles ______________________________
congruent ______________________________
equilateral triangle ______________________________
isosceles triangle ______________________________
obtuse triangle ______________________________
right triangle ______________________________
scalene triangle ______________________________

Directions

1. Cut each strip from the page before writing the definition.
2. Begin the definition on the same line as the word.
3. If a second line is needed, flip the strip toward you and continue on the top line. If a third line is needed, flip the strip back to the original side and continue on the next line. Continue this process until finished.
4. Hold the strip with the original side in view. Bring the two ends toward each other so the labels on the tabs are visible.
5. Flip the tab on the right and place it over tab A such that neither tab is visible.
6. Tape them in place.
7. Use string and the strips to build a Möbius mobile.

Chapter 5
Möbius Mobile

Flip		Tab A
Flip		Tab A
Flip		Tab A
Flip		Tab A
Flip		Tab A
Flip		Tab A
Flip		Tab A
Flip		Tab A

LESSON 6-1

Perimeter & Area of Rectangles & Parallelograms *pp. 280–282*

Vocabulary

perimeter (p. 280) ______________________

____

area (p. 281) ______________________

____

Additional Examples

Example 1

Find the perimeter of each figure.

A.

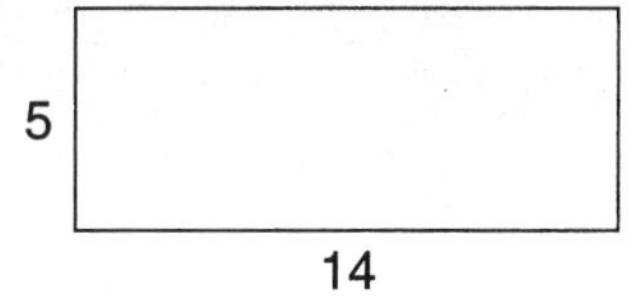

$P =$ ☐ + ☐ + ☐ + ☐ Add all side lengths.

$=$ ☐ units

or $P = 2b + 2h$ Perimeter of rectangle

$= 2($☐$) + 2($☐$)$ Substitute ☐ for b and ☐ for h.

$= 28 + 10 =$ ☐ units

B.

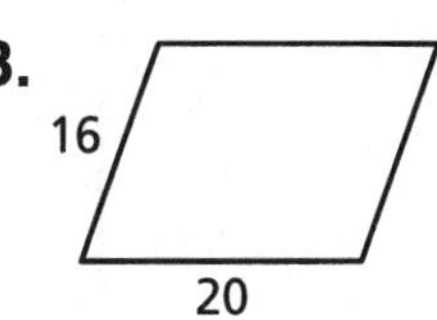

$P =$ ☐ + ☐ + ☐ + ☐

$=$ ☐ units

Example 2

Graph the figure with the given vertices. Then find the area of the figure.

(−1, −2), (2, −2), (2, 3), (−1, 3)

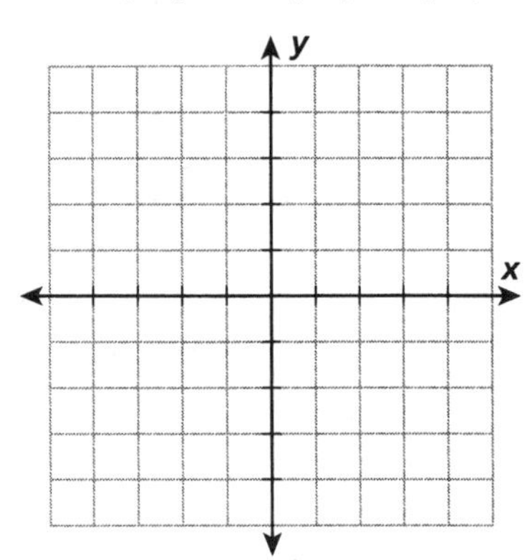

$A = bh$ Area of a rectangle

$=$ $\cdot$ Substitute for b and for h.

$=$ units^2

Try This

1. Find the perimeter of the figure.

6

11

2. Graph the figure with the given vertices. Then find the area of the figure.

(−1, −1), (3, −1), (5, 3), (1, 3)

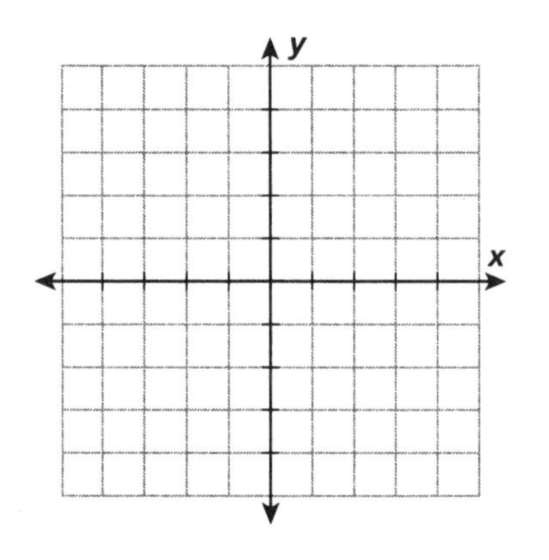

LESSON **6-2** # Perimeter and Area of Triangles and Trapezoids *pp. 285–286*

Additional Examples

Example 1

Find the perimeter of each figure.

A.

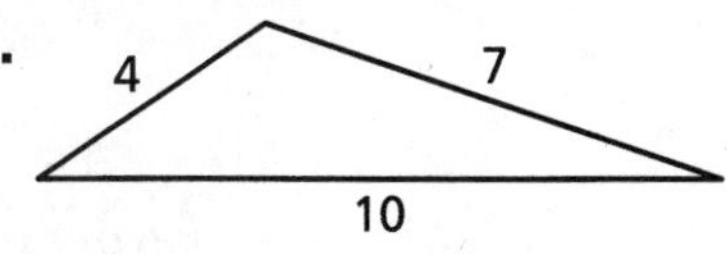

$P =$ ☐ + ☐ + ☐ Add all sides.

$=$ ☐ units

B.

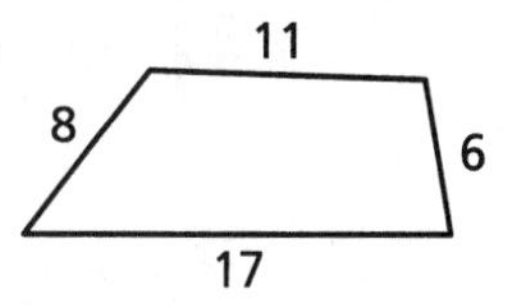

$P =$ ☐ + ☐ + ☐ + ☐ Add all sides.

$=$ ☐ units

Example 2

Graph and find the area of each figure with the given vertices.

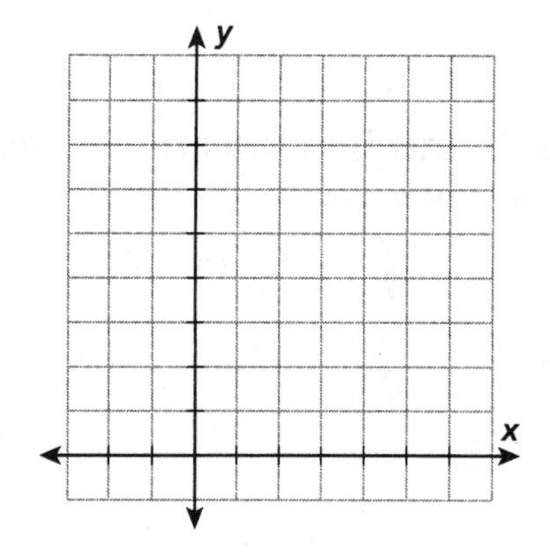

A. (−2, 2), (4, 2), (0, 5)

$A = \frac{1}{2}bh$ ☐ of a triangle

$= \frac{1}{2} \cdot$ ☐ $\cdot$ ☐ Substitute for b and h.

$=$ ☐ units^2

B. (−1, −2), (5, −2), (5, 2), (−1, 6)

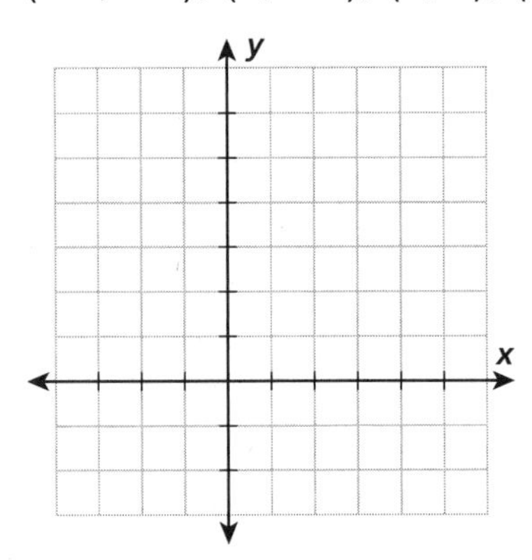

$A = \frac{1}{2}h(b_1 + b_2)$ Area of a

$= \frac{1}{2} \cdot \quad (\quad + \quad)$ Substitute for h, b_1, and b_2.

$=$ units^2

Try This

1. Find the perimeter of the figure.

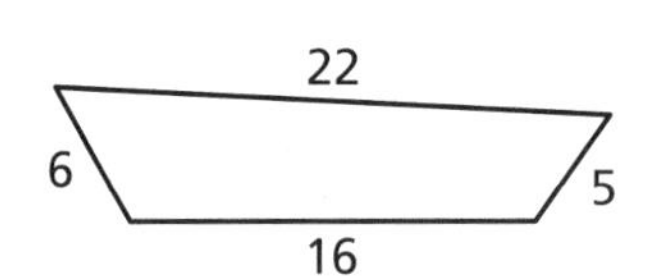

2. Graph and find the area of the figure with the given vertices.

(−2, −1), (0, 5), (3, 5), (5, −1)

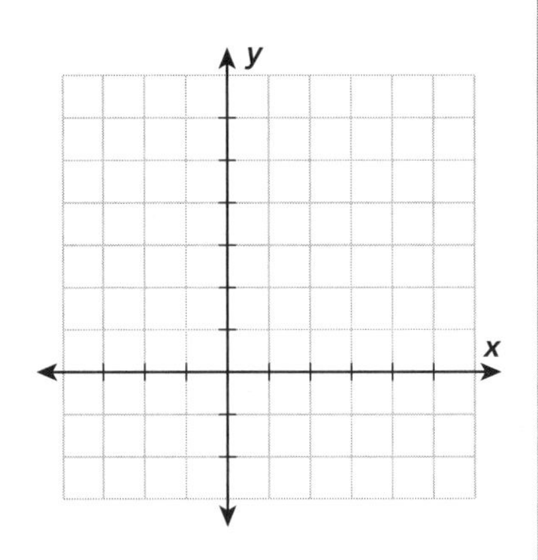

LESSON 6-3

The Pythagorean Theorem

pp. 290–291

Know-It Notes

Vocabulary

Pythagorean Theorem (p. 290) ______

legs (p. 290) ______

hypotenuse (p. 290) ______

Additional Examples

Example 1

Find the length of the hypotenuse.

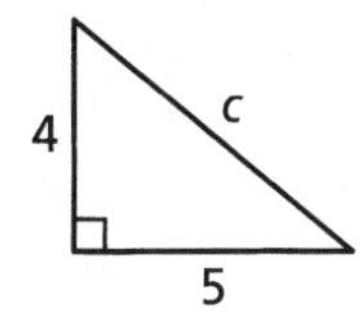

$a^2 + b^2 = c^2$ ______ Theorem

$\square^2 + \square^2 = c^2$ Substitute for a and b.

$\square + \square = c^2$ Simplify powers.

$\square = c^2$

$\square = c$ Solve for c; $c = \sqrt{c^2}$.

$\square \approx c$

Example 2

Solve for the unknown side in the right triangle.

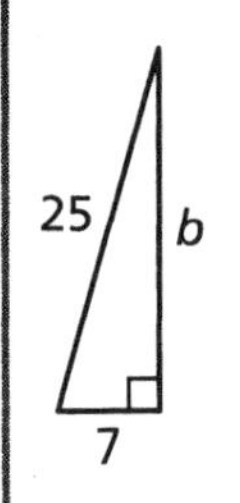

$a^2 + b^2 = c^2$ Theorem

$\square^2 + b^2 =$ Substitute for a and c.

$+ b^2 =$ Simplify powers.

$-$ $-$

$b^2 =$

$b =$ $\sqrt{576} = 24$

Try This

1. Find the length of the hypotenuse.

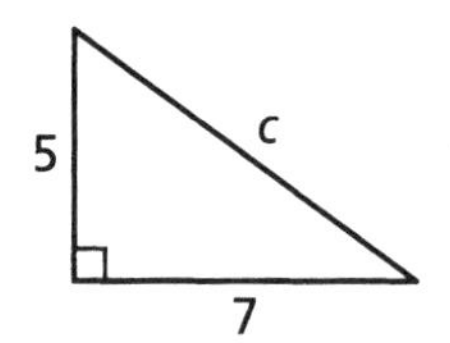

2. Solve for the unknown side in the right triangle.

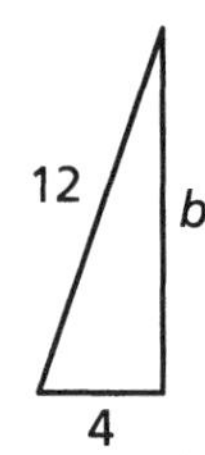

LESSON 6-4

Circles

pp. 294–295

Know-It Notes

Vocabulary

circle (p. 294) ____________________

____

radius (p. 294) ____________________

____

diameter (p. 294) ____________________

____

circumference (p. 294) ____________________

____

Additional Examples

Example 1

Find the circumference of each circle, both in terms of π and to the nearest tenth. Use 3.14 for π.

A. circle with a radius of 4 m

$C = 2\pi r$

$= 2\pi(\square)$

$= \square\,\pi \text{ m} \approx \square \text{ m}$

B. circle with a diameter of 3.3 ft

$C = \pi d$

$= \pi(\square)$

$= \square\,\pi \text{ ft} \approx \square \text{ ft}$

Example 2

Find the area of each circle, both in terms of π and to the nearest tenth. Use 3.14 for π.

A. circle with a radius of 4 in.

$A = \pi r^2 = \pi(\quad)^2$

$= \quad \pi \text{ in}^2 \approx \quad \text{in}^2$

B. circle with a diameter of 3.3 m

$A = \pi r^2 = \pi(\quad)^2 \qquad \frac{d}{2} =$

$= \quad \pi \text{ m}^2 \approx \quad \text{m}^2$

Try This

1. **Find the circumference of the circle, both in terms of π and to the nearest tenth. Use 3.14 for π.**

   circle with a diameter of 4.25 in.

2. **Find the area of each circle, both in terms of π and to the nearest tenth. Use 3.14 for π.**

   circle with a radius of 8 cm

LESSON **6-5**

Drawing Three-Dimensional Figures

pp. 302–303

Know-It Notes

Vocabulary

face (p. 302)

edge (p. 302)

vertex (p. 302)

perspective (p. 303)

vanishing point (p. 303)

horizon line (p. 303)

Additional Examples

Example 1

Use isometric dot paper to sketch a rectangular box that is 5 units long, 3 units deep and 2 units tall.

Step 1: Lightly draw the edges of the bottom face. It will look like a parallelogram.

units by units

Step 2: Lightly draw the vertical line segments from the vertices of the base.

units high

Step 3: Lightly draw the top face by connecting the vertical lines to form a parallelogram.

units by units

Step 4: Darken the lines. Use lines for the edges that are visible and lines for the edges that are hidden.

Example 2

Sketch a one-point perspective drawing of a cube.

Step 1: Draw a square. This will be the front face.

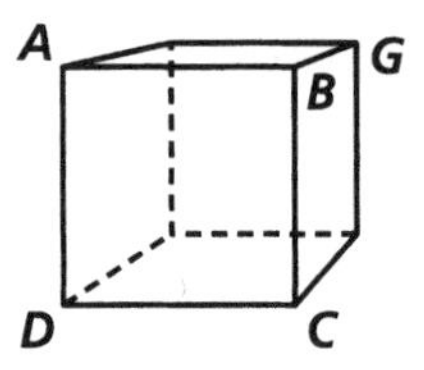

Label the *A* through *D*.

Step 2: Mark a point *V* above your square, and draw a dashed line from each vertex to *V*.

Step 3: Choose a point *G* on $\overline{BV}$ and draw a smaller square that has *G* as one of its vertices.

Step 4: Darken the visible edges, and draw dashed segments for the hidden edges. Erase the point and the lines connecting it to the vertices.

Try This

1. Use isometric dot paper to sketch a rectangular box that is 4 units long, 2 units deep and 3 units tall.

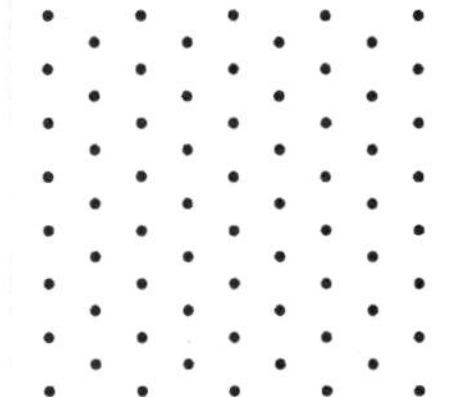

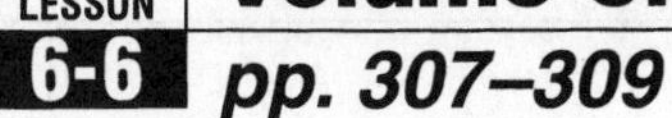

Volume of Prisms and Cylinders

LESSON 6-6 pp. 307–309

Vocabulary

prism (p. 307) ______

cylinder (p. 307) ______

Additional Examples

Example 1

Find the volume of each figure to the nearest tenth.

A. A rectangular prism with base 2 cm by 5 cm and height 3 cm.

$B = \square \cdot \square = \square$ cm^2 Area of $\square$

$V = Bh$ $\square$ of a prism

$= \square \cdot \square$

$= \square$ cm^3

B.

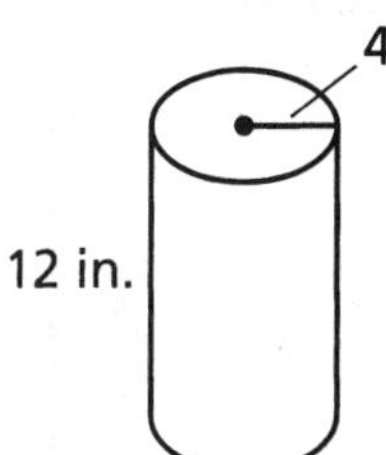

$B = \pi(\square^2) = \square \pi$ in^2 $\square$ of base

$V = Bh$ Volume of a $\square$

$= \square \pi \cdot \square$

$= \square \pi \approx \square$ in^3

Example 4

Find the volume of the barn.

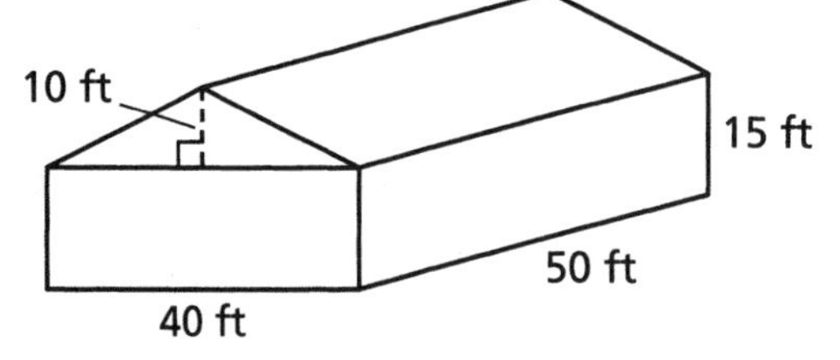

Volume of barn	=	Volume of rectangular prism	+	Volume of triangular prism

$V = (40)(50)(15) \quad + \frac{1}{2}(40)(10)(50)$

$= \qquad\qquad + \qquad$

$= \qquad\qquad$ ft^3

The volume is 40,000 ft^3.

Try This

1. Find the volume of the figure to the nearest tenth.

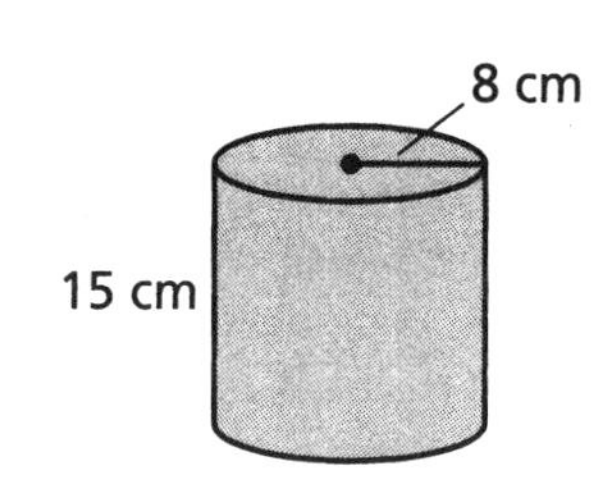

2. Find the volume of the figure.

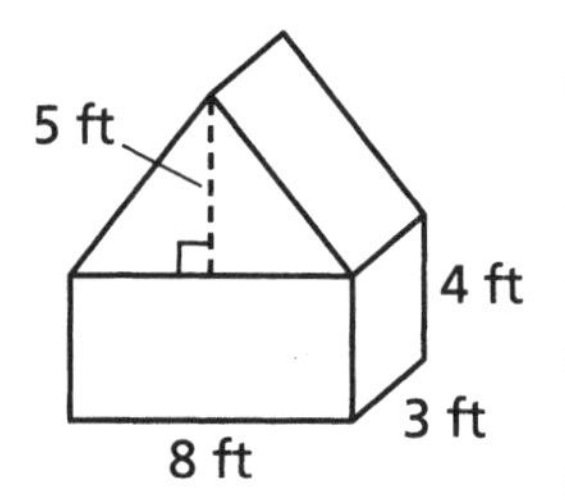

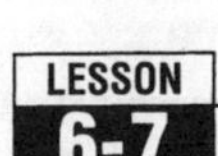

Volume of Pyramids and Cones

pp. 312–313

Vocabulary

pyramid (p. 312) ______________________________

cone (p. 312) ______________________________

Additional Examples

Example 1

Find the volume of each figure.

A.

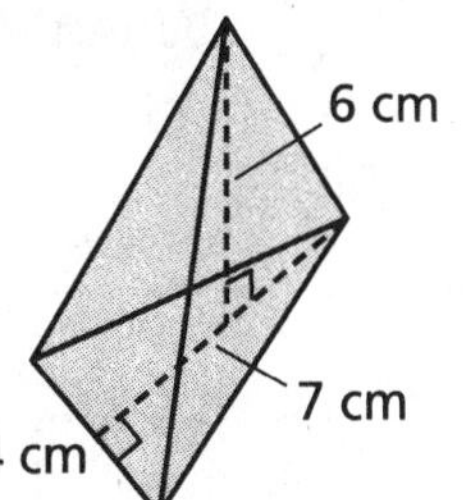

$B = \frac{1}{2}(\square \cdot \square) = \square$ cm²

$V = \frac{1}{3} \cdot \square \cdot \square$ $\qquad V = \frac{1}{3}Bh$

$V = \square$ cm³

B.

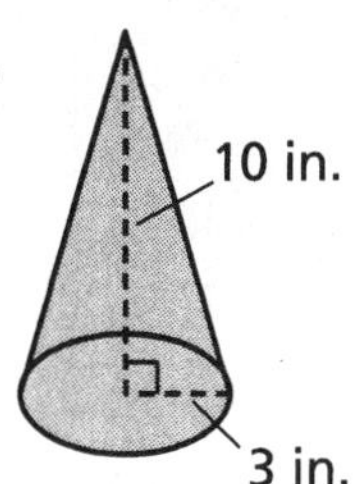

$B = \pi(\square)^2 = \square\pi$ in²

$V = \frac{1}{3} \cdot \square\pi \cdot \square$ $\qquad V = \frac{1}{3}Bh$

$V = \square\pi \approx \square$ in³ $\qquad$ Use 3.14 for $\square$.

Example 2

A cone has a radius of 3 ft. and a height of 4 ft. Explain whether tripling the height would have the same effect on the volume of the cone as tripling the radius.

Original Dimensions	Triple the Height	Triple the Radius
$V = \frac{1}{3}\pi r^2 h$ $= \frac{1}{3}\pi(3^2)4$ $\approx$	$V = \frac{1}{3}\pi r^2(3h)$ $= \frac{1}{3}\pi(3^2)(3 \cdot 4)$ $\approx$	$V = \frac{1}{3}\pi(3r)^2 h$ $= \frac{1}{3}\pi(3 \cdot 3)^2(4)$ $\approx$

When the height of the cone is tripled, the volume is . When the radius is tripled, the volume becomes times the original volume.

Try This

1. Find the volume of the figure.

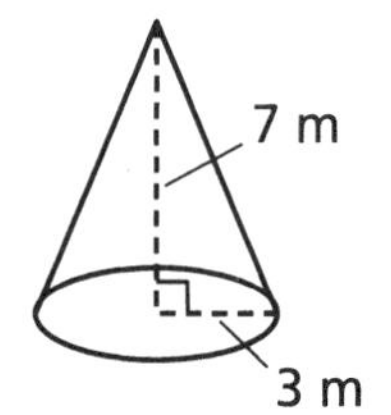

2. A cone has a radius of 2 m and a height of 5 m. Explain whether doubling the height would have the same effect on the volume of the cone as doubling the radius.

LESSON 6-8

Surface Area of Prisms and Cylinders

pp. 316–317

Know-It Notes

Vocabulary

surface area (p. 316) ______

lateral face (p. 316) ______

lateral surface (p. 316) ______

Additional Examples

Example 1

Find the surface area of each figure.

A.

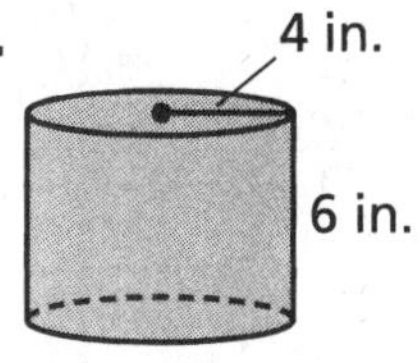

$S = 2\pi r^2 + 2\pi rh$

$= 2\pi(\ \ )^2 + 2\pi(\ \ )(\ \ )$

$= \_\_\_\ \pi\ \text{in}^2 \approx \_\_\_\ \text{in}^2$

B.

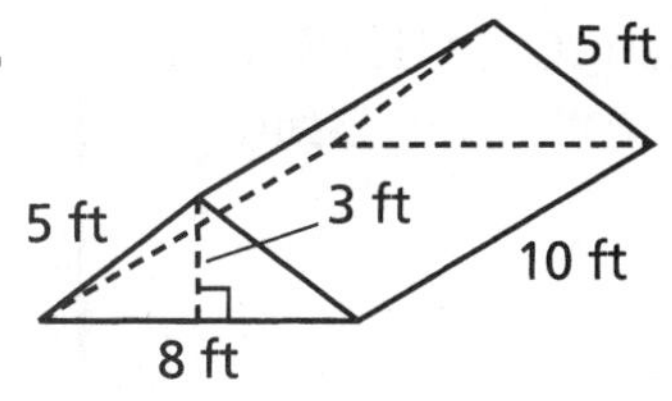

$S = 2B + P = 2(\ \ \cdot\ \ \cdot\ \ ) + (\ \ )(\ \ )$

$= \_\_\_\ \text{ft}^2$

Example 2

A cylinder has diameter 8 in. and height 3 in. Explain whether tripling the height would have the same effect on the surface area as tripling the radius.

Original Dimensions	Triple the Height	Triple the Radius
$S = 2\pi r^2 + 2\pi rh$ $= 2\pi(4)^2 + 2\pi(4)(3)$ $= 56\pi \text{ in}^2 \approx$	$S = 2\pi r^2 + 2\pi r(3h)$ $= 2\pi(4)^2 + 2\pi(4)(9)$ $= 104\pi \text{ in}^2 \approx$	$S = 2\pi(3r)^2 + 2\pi(3r)h$ $= 2\pi(12)^2 + 2\pi(12)(3)$ $= 360\pi \text{ in}^2 \approx$

They   have the same effect. Tripling the radius would increase the surface area   than tripling the height.

Try This

1. Find the surface area of the figure.

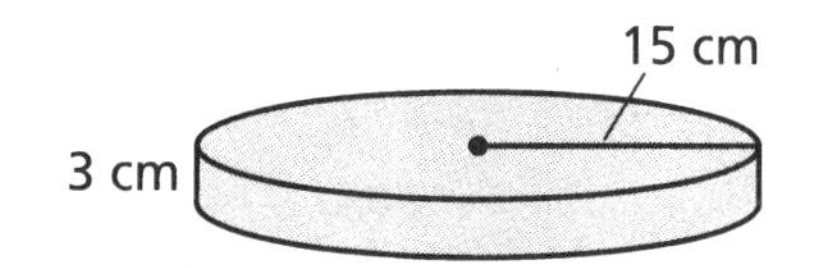

2. A cylinder has diameter 6 in. and height 2 in. Explain whether doubling the height would have the same effect on the surface area as doubling the radius.

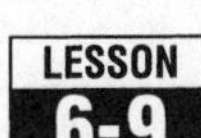

Surface Area of Pyramids and Cones

pp. 320–321

Know-It Notes

Vocabulary

slant height (p. 320) ____________________

regular pyramid (p. 320) ____________________

right cone (p. 320) ____________________

Additional Examples

Example 1

Find the surface area of each figure.

A.

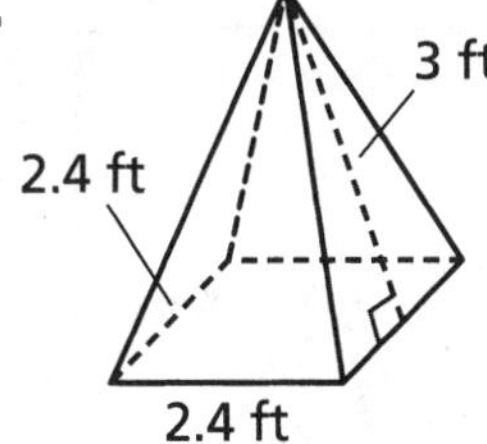

$S = B + \frac{1}{2}Pl$

$= (\quad \cdot \quad) + \frac{1}{2}(\quad)(\quad)$

$= \quad$ ft²

B.

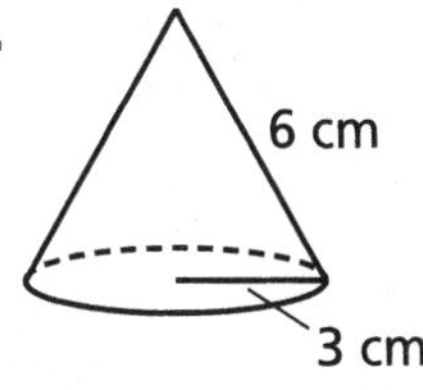

$S = \pi r^2 + \pi r l$

$= \pi(\quad)^2 + \pi(\quad)(\quad)$

$= \quad \pi \approx \quad$ cm²

Example 2

A cone has diameter 8 in. and slant height 3 in. Explain whether tripling the slant height would have the same effect on the surface area as tripling the radius.

Original Dimensions	Triple the Slant Height	Triple the Radius
$S = \pi r^2 + \pi rl$ $= \pi(4)^2 + \pi(4)(3)$ $= 28\pi \text{ in}^2 \approx$	$S = \pi r^2 + \pi r(3l)$ $= \pi(4)^2 + \pi(4)(9)$ $= 52\pi \text{ in}^2 \approx$	$S = \pi(3r)^2 + \pi(3r)l$ $= \pi(12)^2 + \pi(12)(3)$ $= 180\pi \text{ in}^2 \approx$

They have the same effect. Tripling the radius would

increase the surface area than tripling the slant height.

Try This

1. Find the surface area of the figure.

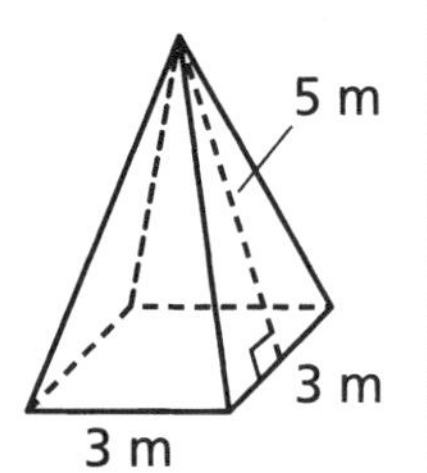

2. A cone has diameter 9 in. and a slant height 2 in. Explain whether tripling the slant height would have the same effect on the surface area as tripling the radius.

LESSON 6-10

Spheres

pp. 324–325

Vocabulary

sphere (p. 324) ______

hemisphere (p. 324) ______

great circle (p. 324) ______

Additional Examples

Example 1

Find the volume of a sphere with radius 9 cm, both in terms of π and to the nearest tenth of a unit.

$V = \left(\frac{4}{3}\right)\pi r^3$ ______ of a sphere

$= \left(\frac{4}{3}\right)\pi(\_\_\_^3)$ Substitute ______ for r.

$=$ ______ π cm$^3 \approx$ ______ cm^3

Example 2

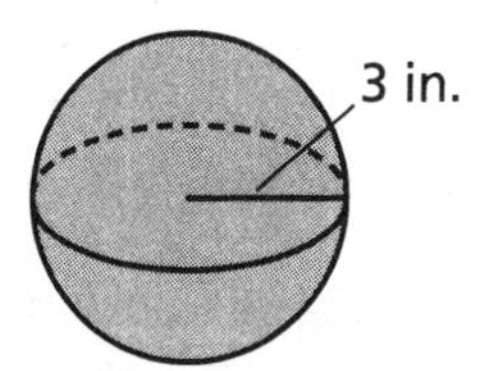

Find the surface area, both in terms of π and to the nearest tenth of a unit.

$S = 4\pi r^2$ ______ area of a sphere

$= 4\pi(\_\_\_^2)$ Substitute ______ for r.

$=$ ______ π in$^2 \approx$ ______ in^2

Example 3

Compare the volume and surface area of a sphere with radius 42 cm with those of a rectangular prism measuring 44 × 84 × 84 cm.

Sphere:

$V = \left(\frac{4}{3}\right)\pi r^3 = \left(\frac{4}{3}\right)\pi(\quad)^3$

$\approx \left(\frac{4}{3}\right)\left(\frac{22}{7}\right)$

$\approx$ cm^3

$S = 4\pi r^2 = 4\pi(\quad)^2$

$= \quad \pi$

$\approx \quad \left(\frac{22}{7}\right)$

$\approx$ cm^2

Rectangular Prism:

$V = lwh$

$= (\quad)(\quad)(\quad)$

$=$ cm^3

$S = 2lw + 2lh + 2wh$

$= 2(\quad)(\quad) +$

$2(\quad)(\quad) +$

$2(\quad)(\quad)$

$=$ cm^2

The sphere and the prism have approximately the same ,

but the prism has a surface area.

Try This

1. **Find the volume of a sphere with radius 3 cm, both in terms of π and to the nearest tenth of a unit.**

2. **The moon has a radius of 1,738 km. Find the surface area, both in terms of π and to the nearest tenth.**

Chapter 6

Vocabulary Chain

EXAMPLE		TERM
		circumference
EXAMPLE		TERM
		perimeter
EXAMPLE		TERM
		hypotenuse
EXAMPLE		TERM
		Pythagorean Theorem
EXAMPLE		TERM
		surface area

Directions

1. Write an example and an explanation in words, numbers, and algebra for each term.
2. Cut out each vocabulary strip and fold in thirds.
3. Punch a hole in the corner of each folded vocabulary strip, and string them together to create your vocabulary chain.

Chapter 6
Vocabulary Chain

WORDS	NUMBERS	ALGEBRA
WORDS	NUMBERS	ALGEBRA
WORDS	NUMBERS	ALGEBRA
WORDS	NUMBERS	ALGEBRA
WORDS	NUMBERS	ALGEBRA

LESSON **7-1**

Ratios and Proportions

pp. 342–343

Vocabulary

ratio (p. 342) ______________________

equivalent ratios (p. 342) ______________________

proportion (p. 343) ______________________

Additional Examples

Example 1

Find two ratios that are equivalent to each given ratio.

A. $\frac{9}{27}$

$\frac{9}{27} = \frac{9 \cdot 2}{27 \cdot 2} = \square$

$\frac{9}{27} = \frac{9 \div 9}{27 \div 9} = \square$

Multiply or divide the ______ and ______ by the same nonzero number.

Two ratios equivalent to $\frac{9}{27}$ are ______ and ______.

B. $\frac{64}{24}$

$\frac{64}{24} = \frac{64 \cdot 2}{24 \cdot 2} = \square$

$\frac{64}{24} = \frac{64 \div 8}{24 \div 8} = \square$

Two ratios equivalent to $\frac{64}{24}$ are ______ and ______.

Example 2

Simplify to tell whether the ratios form a proportion.

A. $\frac{3}{27}$ and $\frac{2}{18}$

$\frac{3}{27} = \frac{3 \div \quad}{27 \div \quad} =$

$\frac{2}{18} = \frac{2 \div \quad}{18 \div \quad} =$

Since $\quad = \quad$, the ratios in proportion.

B. $\frac{12}{15}$ and $\frac{27}{36}$

$\frac{12}{15} = \frac{12 \div \quad}{15 \div \quad} =$

$\frac{27}{36} = \frac{27 \div \quad}{36 \div \quad} =$

Since $\quad \neq \quad$, ratios in proportion.

Try This

1. Find two ratios that are equivalent to the given ratio.

$\frac{8}{16}$

2. Simplify to tell whether the ratios form a proportion.

$\frac{14}{49}$ and $\frac{16}{36}$

LESSON 7-2

Ratios, Rates, and Unit Rates

pp. 346–347

Vocabulary

rate (p. 346) ______

unit rate (p. 346) ______

unit price (p. 347) ______

Additional Examples

Example 1

Order the ratios 4:3, 23:10, 13:9, and 47:20 from the least to greatest.

A. 4:3 = $\frac{\square}{\square}$ = ______ Divide. $\frac{4}{3} = \frac{1.\overline{3}}{1}$

23:10 = $\frac{\square}{\square}$ = ______

13:9 = $\frac{\square}{\square}$ = ______

47:20 = $\frac{\square}{\square}$ = ______

The decimals in order are ______, ______, ______, and ______.

The ratios in order from least to greatest are ______, ______, ______, and ______.

Example 2

Use the bar graph to find the number of acres, to the nearest acre, destroyed in Nevada and Alaska per week.

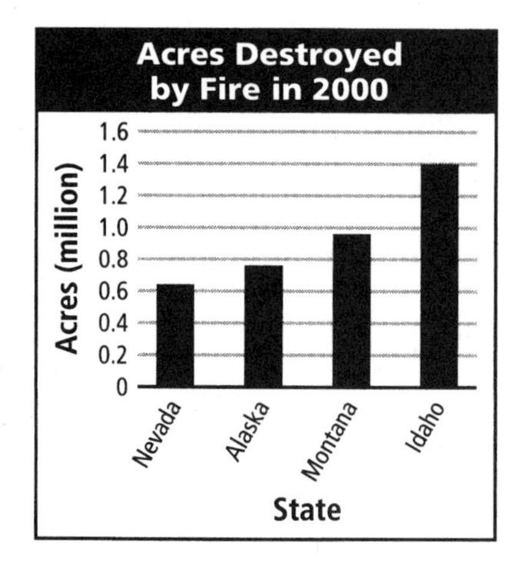

Nevada = $\dfrac{\qquad\text{acres}}{\qquad\text{weeks}} \approx$

$\approx$

Alaska = $\dfrac{\qquad\text{acres}}{\qquad\text{weeks}} \approx$

Try This

1. Order the ratios 2:3, 35:14, 5:3, and 49:20 from the least to greatest.

2. Use the above bar graph to find the number of acres, to the nearest acre, destroyed in Montana and Idaho per week.

LESSON 7-3

Analyze Units

pp. 350–352

Vocabulary

conversion factor (p. 350) ____________________

____

____

Additional Examples

Example 3

PROBLEM SOLVING APPLICATION

A car traveled 60 miles on a road in 2 hours. How many feet per second was the car traveling?

1. Understand the Problem

The problem is stated in units of miles and hours. The question asks for the answer in units of feet and seconds. You will need to use several conversion factors.

List the important information:

- Miles to feet → $\frac{\_\_\_\_ \text{ ft}}{\_\_\_\_ \text{ mi}}$
- Hours to minutes → $\frac{\_\_\_\_ \text{ h}}{\_\_\_\_ \text{ min}}$
- Minutes to seconds → $\frac{\_\_\_\_ \text{ min}}{\_\_\_\_ \text{ s}}$

2. Make a Plan

Multiply by each ____________ factor separately, or simplify the problem and multiply by several ____________ factors at once.

3. Solve

First, convert 60 miles in 2 hours into a unit rate.

$$\frac{60\text{ mi}}{2\text{ h}} = \frac{(60 \div 2)\text{ mi}}{(2 \div 2)\text{ h}} = \frac{\_\_\_\_\_\_\text{ mi}}{\_\_\_\_\_\_\text{ h}}$$

Create a single conversion factor to convert hours directly to seconds:

hours to minutes → $\frac{\_\_\_\_\_\_\text{ h}}{\_\_\_\_\_\_\text{ min}}$; minutes to seconds → $\frac{\_\_\_\_\_\_\text{ min}}{\_\_\_\_\_\_\text{ s}}$

hours to seconds $= \frac{1\text{ h}}{60\text{ }\cancel{\text{min}}} \cdot \frac{1\text{ }\cancel{\text{min}}}{60\text{ s}} = \frac{\_\_\_\_\_\_\text{ h}}{\_\_\_\_\_\_\text{ s}}$

$\frac{30\text{ mi}}{1\text{h}} \cdot \frac{5280\text{ ft}}{1\text{ mi}} \cdot \frac{1\text{ h}}{3600\text{ s}}$ Set up the factors.

Do not include the numbers yet. Notice what happens to the units.

$\frac{\text{mi}}{\text{h}} \cdot \frac{\text{ft}}{\text{mi}} \cdot \frac{\text{h}}{\text{s}}$ Simplify. Only remains.

$\frac{30\text{ mi}}{1\text{h}} \cdot \frac{5280\text{ ft}}{1\text{ mi}} \cdot \frac{1\text{ h}}{3600\text{ s}}$ Multiply.

$\frac{30 \cdot 5280\text{ ft} \cdot 1}{1 \cdot 1 \cdot 3600\text{ s}} = \frac{158{,}400\text{ ft}}{1\text{ s}} = \frac{\_\_\_\_\_\_\text{ ft}}{\_\_\_\_\_\_\text{ s}}$ Multiply.

The car was traveling feet per second.

4. Look Back

A rate of ft/s is less than 50 ft/s. A rate of 60 miles in 2 hours is 30 mi/h or 0.5 mi/min.

Since 0.5 mi/min is less than 3000 ft/60 s or 50 ft/s and 44 ft/s is less than 50 ft/s, then 44 ft/s is a reasonable answer.

Try This

1. Problem Solving Application

A train traveled 180 miles on a railroad track in 4 hours. How many feet per second was the train traveling?

1. Understand the Problem

The problem is stated in units of miles and hours. The question asks for the answer in units of feet and seconds. You will need to use several conversion factors.

List the important information:

Miles to feet → $\frac{\square \text{ ft}}{\square \text{ mi}}$

Hours to minutes → $\frac{\square \text{ h}}{\square \text{ min}}$

Minutes to seconds → $\frac{\square \text{ min}}{\square \text{ s}}$

2. Make a Plan

Multiply by each ____________ factor separately, or simplify the problem and multiply by several ____________ factors at once.

3. Solve

4. Look Back

A rate of 66 ft/s is more than 50 ft/s. A rate of 180 miles in 4 hours is 45 mi/h or 0.75 mi/min.

Since 0.75 mi/min is more than 3000 ft/60 s or 50 ft/s and 66 ft/s is more than 50 ft/s, then 66 ft/s is a reasonable answer.

LESSON 7-4

Solving Proportions
pp. 356–357

Vocabulary

cross product (p. 356)

Additional Examples

Example 1

Tell whether the ratios are proportional.

A. $\frac{6}{15} \stackrel{?}{=} \frac{4}{10}$

$\frac{6}{15} \times \frac{4}{10}$ → 60, 60   Find products.

=

Since the cross products are , the ratios proportional.

B. A mixture of fuel for a certain small engine should be 4 parts gasoline to 1 part oil. If you combine 5 quarts of oil with 15 quarts of gasoline, will the mixture be correct?

$\frac{\text{4 parts gasoline}}{\text{1 part oil}} \stackrel{?}{=} \frac{\text{15 quarts gasoline}}{\text{5 quarts oil}}$ Set up ratios.

• = 20   • =   Find the cross .

≠

The ratios equal. The mixture correct.

Example 3

Allyson weighs 55 lbs and sits on a seesaw 4 ft away from its center. If Marco sits 5 ft away from the center and the seesaw is balanced, how much does Marco weigh?

$\frac{\text{mass 1}}{\text{length 2}} = \frac{\text{mass 2}}{\text{length 1}}$ Set up the ______.

$\frac{\square}{\square} = \frac{\square}{\square}$ Let x represent Marco's weight.

$\square \cdot \square = \square x$ Find the cross products.

$\square = \square x$ Multiply.

$\frac{220}{\square} = \frac{5x}{\square}$ Solve. Divide both sides by ___.

$\square = x$

Marco weighs ___ lb.

Try This

1. Tell whether the ratios are proportional.

$\frac{5}{10} \stackrel{?}{=} \frac{2}{4}$

2. Solve the proportion.

$\frac{14}{g} = \frac{2}{3}$

LESSON **7-5** **Dilations**

pp. 362–363

Vocabulary

dilation (p. 362)

scale factor (p. 362)

center of dilation (p. 362)

Additional Examples

Example 1

Tell whether each transformation is a dilation.

A.

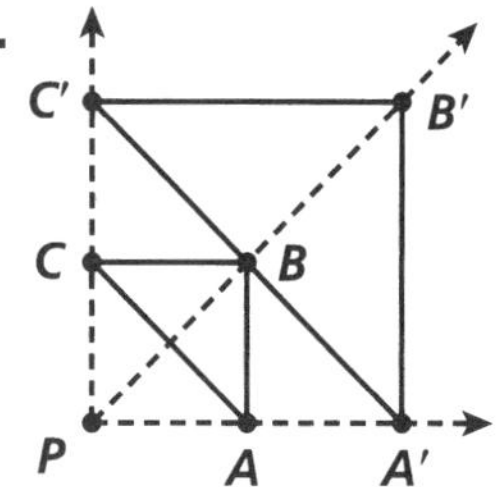

The transformation

a dilation.

B.

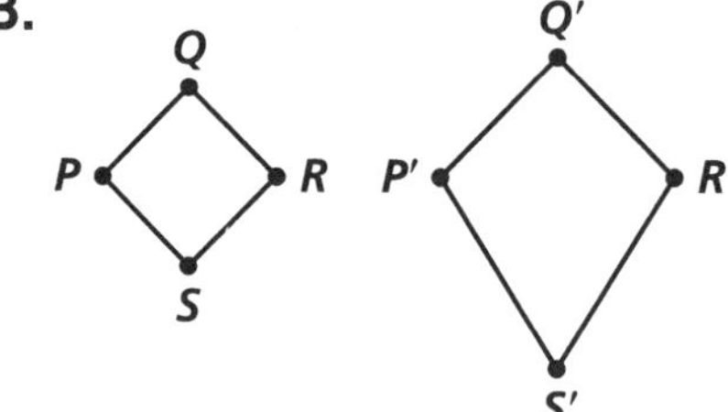

The transformation a

dilation. The figure is distorted.

C.

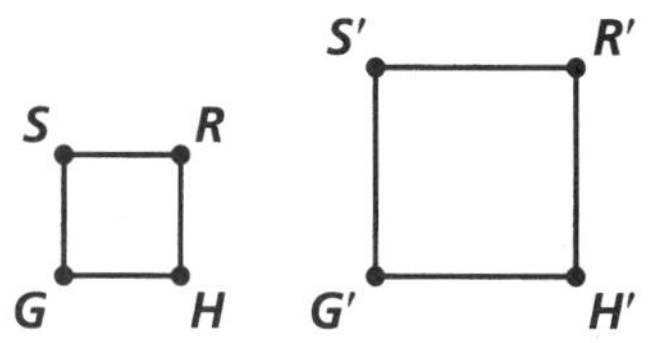

The transformation

a dilation.

D.

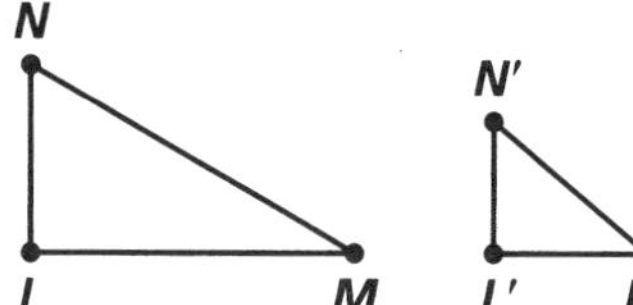

The transformation a

dilation. The figure is distorted.

Example 2

Dilate the figure by a scale factor of 1.5 with *P* as the center of dilation.

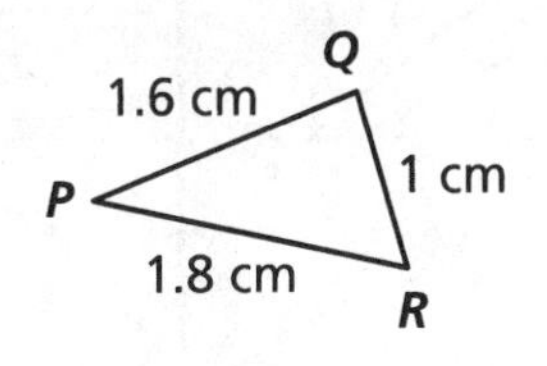

Multiply each side by ______.

Example 3

A. Use the origin as the center of dilation and dilate the figure in Example 3A on page 363 by a scale factor of 2. What are the vertices of the image?

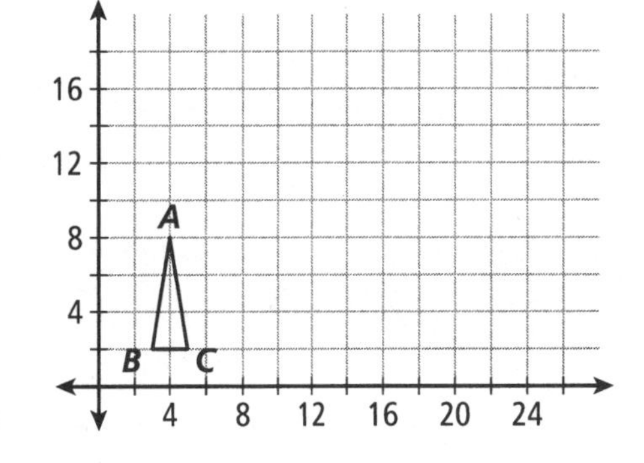

Multiply the coordinates by ______ to find the vertices of the image.

$\triangle ABC$ $\qquad\qquad$ $\triangle A'B'C'$

$A(4, 8) \rightarrow A'(4 \cdot$ ______ $, 8 \cdot$ ______ $) \rightarrow A'$ (______ , ______)

$B(3, 2) \rightarrow B'(3 \cdot$ ______ $, 2 \cdot$ ______ $) \rightarrow B'$ (______ , ______)

$C(5, 2) \rightarrow C'(5 \cdot$ ______ $, 2 \cdot$ ______ $) \rightarrow C'$ (______ , ______)

The vertices of the image are A' (______ , ______), B' (______ , ______), and C' (______ , ______).

Try This

1. Tell whether the transformation is a dilation.

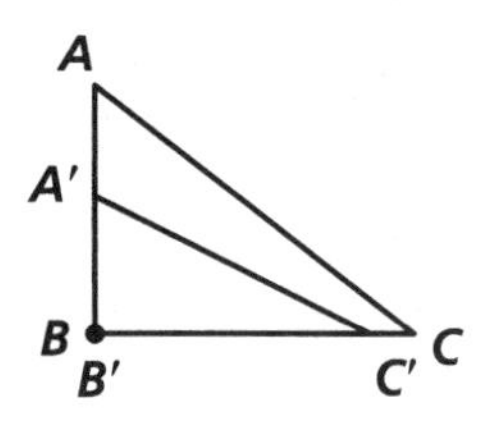

2. Dilate the figure by a scale factor of 0.5 with G as the center of dilation.

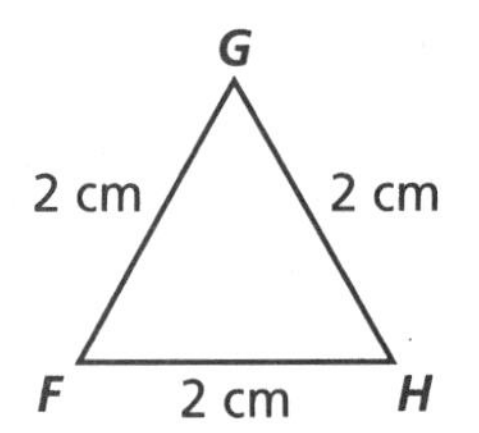

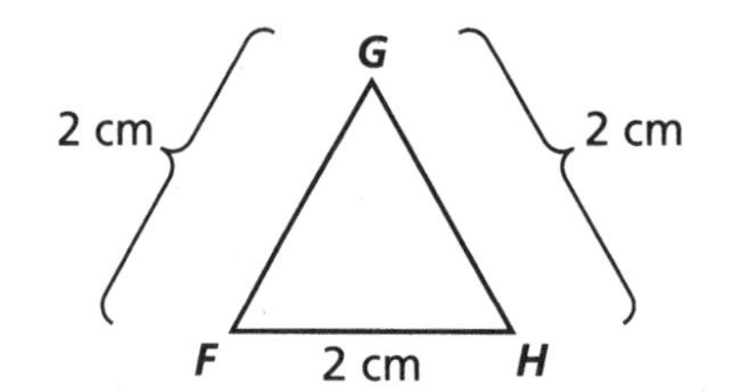

3. Use the origin as the center of dilation and dilate the figure by a scale factor of 2. What are the vertices of the image?

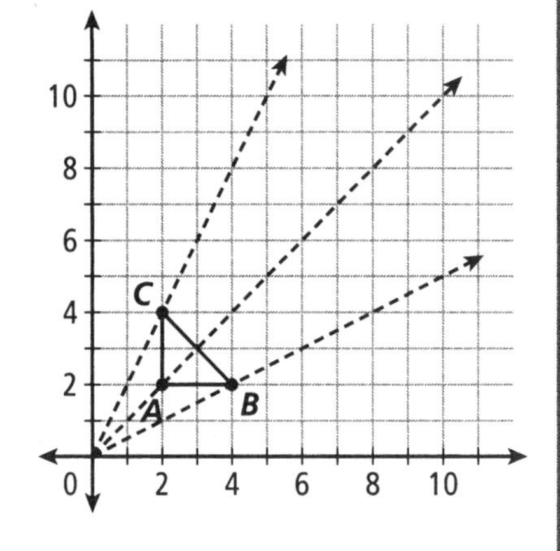

(,) (,)
(,)

LESSON **7-6**

Similar Figures

pp. 368–369

Vocabulary

similar (p. 368)

Additional Examples

Example 1

A picture 10 in. tall and 14 in. wide is to be scaled to 1.5 in. tall to be displayed on a Web page. How wide should the picture be on the Web page for the two pictures to be similar?

To find the ______ factor, divide the known measurement of the scaled picture by the corresponding measurement of the original picture.

0.15 $\frac{1.5}{10} =$ ______

Then multiply the width of the original picture by the scale factor.

2.1 $14 \cdot 0.15 =$ ______

The picture should be ______ in. wide.

Example 2

A T-shirt design includes an isosceles triangle with side lengths 4.5 in, 4.5 in., and 6 in. An advertisement shows an enlarged version of the triangle with two sides that are each 3 ft. long. What is the length of the third side of the triangle in the advertisement?

Set up a proportion.

$\frac{\underline{\quad\quad} \text{ in.}}{\quad \text{ft}} = \frac{\underline{\quad\quad} \text{ in.}}{\quad \text{ft}}$.

$4.5 \text{ in.} \cdot x \text{ ft} = 3 \text{ ft} \cdot 6 \text{ in.}$	Find the products.
$4.5 \text{ in.} \cdot x \text{ ft} = 3 \text{ ft} \cdot 6 \text{ in.}$	in. • ft is on both sides
$4.5x = 3 \cdot 6$	Cancel the units.
$4.5x = 18$	Multiply.
$x = \quad = $	Solve for x.

The third side of the triangle is ft long.

Try This

1. **A painting 40 in. tall and 56 in. wide is to be scaled to 10 in. tall to be displayed on a poster. How wide should the painting be on the poster for the two pictures to be similar?**

2. **A flag in the shape of an isosceles triangle with side lengths 18 ft, 18 ft, and 24 ft is hanging on a pole outside a campground. A camp t-shirt shows a smaller version of the triangle with two sides that are each 4 in. long. What is the length of the third side of the triangle on the t-shirt?**

LESSON 7-7

Scale Drawings

pp. 372–373

Know-It Notes

Vocabulary

scale drawing (p. 372) ______________________________

scale (p. 372) ______________________________

reduction (p. 373) ______________________________

enlargement (p. 373) ______________________________

Additional Examples

Example 1

A. The length of an object on a scale drawing is 2 cm, and its actual length is 8 m. The scale is 1 cm: __ m. What is the scale?

$\frac{1 \text{ cm}}{x \text{ m}} = \frac{\square \text{ cm}}{\square \text{ m}}$ Set up proportion using $\frac{\text{scale length}}{\text{actual length}}$.

$1 \cdot \square = x \cdot \square$ Find the cross products.

$\square = 2x$

$\square = x$ Solve the proportion.

The scale is 1 cm: $\square$ m.

Example 3

A. If a wall in a $\frac{1}{4}$ in. scale drawing is 4 in. tall, how tall is the actual wall?

$\frac{0.25 \text{ in.}}{1 \text{ ft}} = \frac{4 \text{ in.}}{x \text{ ft}}$ ← scale length / ← actual length — Length ratios are equal.

$\cdot\ x = \quad \cdot$ — Find the cross products.

$x =$ — Solve the .

The wall is ft tall.

B. How tall is the wall if a $\frac{1}{2}$ in. scale is used?

$\frac{0.5 \text{ in.}}{1 \text{ ft}} = \frac{4 \text{ in.}}{x \text{ ft}}$ ← scale length / ← actual length — Length ratios are equal.

$\cdot\ x = \quad \cdot$ — Find the cross .

$x =$ — Solve the .

The wall is ft tall.

Try This

1. **The length of an object on a scale drawing is 4 cm, and its actual length is 12 m. The scale is 1 cm: __ m. What is the scale?**

2. **If a wall in a $\frac{1}{4}$ in. scale drawing is 0.5 in. thick, how thick is the actual wall?**

LESSON 7-8

Scale Models

pp. 376–377

Vocabulary

scale model (p. 376) ______

Additional Examples

Example 1

Tell whether each scale reduces, enlarges, or preserves the size of the actual object.

A. 1 in:1 yd

$\frac{1 \text{ in.}}{1 \text{ yd}} = \frac{1 \text{ in.}}{\square \text{ in.}} = \square$ Convert: 1 yd = 36 in. Simplify.

The scale ______ the size of the actual object by a factor of ______.

B. 1 m:10 cm

$\frac{1 \text{ m}}{10 \text{ cm}} = \frac{\square \text{ cm}}{10 \text{ cm}} = 10$ Convert: 1 m = 100 cm. Simplify.

The scale ______ the size of the actual object 10 times.

Example 2

What scale factor relates a 12 in. scale model to a 6 ft. man?

12 in:6 ft State the ______.

$\frac{12 \text{ in.}}{6 \text{ ft}} = \frac{12 \text{ in.}}{\square \text{ in.}} = \square$ Write the scale ______ as a ratio and simplify.

The scale factor is ______, or ______.

Example 3

A model of a 32 ft tall house was made using the scale 3 in:2 ft. What is the height of the model?

$\frac{3\text{ in.}}{2\text{ ft}} = \frac{3\text{ in.}}{\quad\text{ in.}} = \frac{1\text{ in.}}{8\text{ in.}} =$ First find the factor.

The scale factor for the model is . Now set up a proportion.

$\frac{1}{8} = \frac{h\text{ in.}}{384\text{ in.}}$ Convert: 32 ft = 384 in.

$= 8h$ Cross multiply.

$= h$ Solve for the height.

The height of the model is in.

Try This

1. **Tell whether the scale reduces, enlarges, or preserves the size of the actual object.**

   1 in:1 ft

2. **What scale factor relates a 12 in. scale model to a 4 ft. tree?**

3. **A model of 24 ft tall bridge was made using the scale 4 in:2 ft. What is the height of the model?**

LESSON 7-9

Scaling Three-Dimensional Figures

pp. 382–383

Know-It Notes

Vocabulary

capacity (p. 382) ______________________

Additional Examples

Example 1

A 3 cm cube is built from small cubes, each 1 cm on an edge. Compare the following values.

A. the edge lengths of the large and small cubes

$\frac{3 \text{ cm cube}}{1 \text{ cm cube}} \rightarrow \frac{\square \text{ cm}}{\square \text{ cm}} = \square$ Ratio of corresponding ______

The edges of the large cube are ____ times as long as the edges of the small cube.

B. the surface areas of the two cubes

$\frac{3 \text{ cm cube}}{1 \text{ cm cube}} \rightarrow \frac{\square \text{ cm}^2}{\square \text{ cm}^2} = \square$ Ratio of corresponding ______

The surface area of the large cube is ____ times that of the small cube.

C. the volumes of the two cubes

$\frac{3 \text{ cm cube}}{1 \text{ cm cube}} \rightarrow \frac{27 \text{ cm}^3}{1 \text{ cm}^3} = \square$ Ratio of corresponding ______

The volume of the large cube is ____ times that of the small cube.

Example 2

A box is in the shape of a rectangular prism. The box is 4 ft tall, and its base has a length of 3 ft and a width of 2 ft. For a 6 in. tall model of the box, find the following.

A. What is the scale factor of the model?

$\frac{6 \text{ in.}}{4 \text{ ft}} = \frac{6 \text{ in.}}{\quad \text{ in.}} =$ Convert and simplify.

The scale factor of the model is .

B. What are the length and the width of the model?

Length: $\cdot\ 3 \text{ ft} = \frac{36}{8} \text{ in.} =$ in.

Width: $\cdot\ 2 \text{ ft} = \frac{24}{8} \text{ in.} =$ in.

The length of the model is in., and the width is in.

Try This

1. **A 2 cm cube is built from small cubes, each 1 cm on an edge. Compare the following values.**

   the edge lengths of the large and small cubes

2. **A box is in the shape of a rectangular prism. The box is 8 ft tall, and its base has a length of 6 ft and a width of 4 ft. For a 6 in. tall model of the box, find the following.**

   What is the scale factor of the model?

Chapter 7
Möbius Mobile

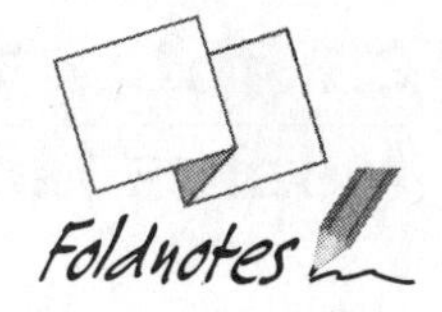

conversion factor ______________________________

dilation ______________________________

proportion ______________________________

rate ______________________________

scale ______________________________

scale factor ______________________________

unit price ______________________________

unit rate ______________________________

Directions

1. Cut each strip from the page before writing the definition.
2. Begin the definition on the same line as the word.
3. If a second line is needed, flip the strip toward you and continue on the top line. If a third line is needed, flip the strip back to the original side and continue on the next line. Continue this process until finished.
4. Hold the strip with the original side in view. Bring the two ends toward each other so the labels on the tabs are visible.
5. Flip the tab on the right and place it over tab A such that neither tab is visible.
6. Tape them in place.
7. Use string and the strips to build a Möbius mobile.

Chapter 7
Möbius Mobile

Flip		Tab A
Flip		Tab A
Flip		Tab A
Flip		Tab A
Flip		Tab A
Flip		Tab A
Flip		Tab A
Flip		Tab A

LESSON 8-1 Relating Decimals, Fractions, and Percents *pp. 400–401*

Know-It Notes

Vocabulary

percent (p. 400) ______________________________

____

Additional Examples

Example 1

Find the missing ratio or percent equivalent for each letter *a–g* on the number line.

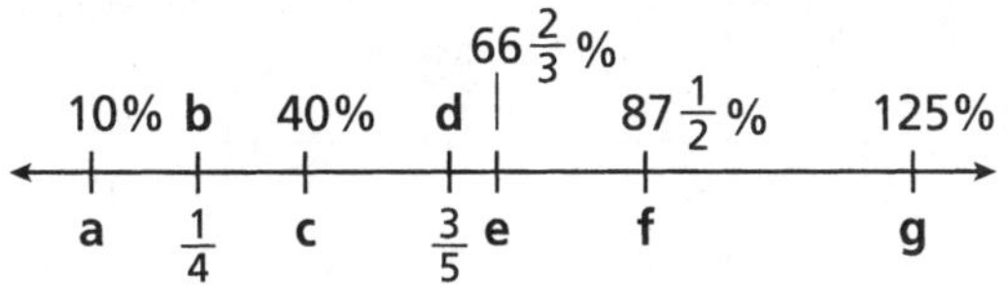

a: $10\% = \frac{\square}{100} = \square$

b: $\frac{1}{4} = 0.\square = \square\%$

c: $40\% = \frac{\square}{100} = \frac{\square}{10} = \square$

d: $\frac{3}{5} = 0.\square = \square\%$

e: $66\frac{2}{3}\% = 0.\square = \square$

f: $87\frac{1}{2}\% = 0.\square = \frac{\square}{1000} = \square$

g: $125\% = \square = \square = \square$

Example 2

Find the equivalent fraction, decimal, or percent for each value given on the circle graph.

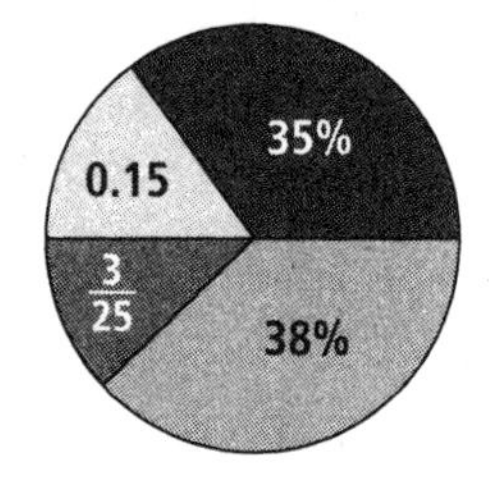

Fraction	Decimal	Percent
$\frac{\quad}{100} =$	0.15	$0.15(100) = \quad \%$
$\frac{\quad}{100} =$	$\frac{7}{20} =$	35%
$\frac{\quad}{100} =$	$\frac{19}{50} =$	38%
$\frac{3}{25}$	$\frac{3}{25} =$	$0.12(100) = \quad \%$

Try This

1. Find the missing ratio or percent equivalent for each letter *a–g* on the number line.

$12\frac{1}{2}\%$ 25% c 50% e 75% g

a b $\frac{3}{8}$ d $\frac{5}{8}$ f 1

2. Fill in the missing pieces on the chart below.

Fraction	Decimal	Percent
	0.1	
		45%
$\frac{1}{4}$		
$\frac{1}{5}$		

LESSON 8-2
Finding Percents
pp. 405–406

Additional Examples

Example 1

A. What percent of 220 is 88?

Method 1: Set up an equation to find the percent.

$p \cdot$ ____ = ____ Set up an equation.

$p = \frac{\_\_\_\_}{\_\_\_\_}$ Solve for p.

$p =$ ____ 0.4 is ____ %.

So 88 is ____ % of 220.

B. Eddie weighs 160 lb, and his bones weigh 24 lb. Find the percent of his weight that his bones are.

Method 2: Set up a proportion to find the percent.

Think: What number is to 100 as ____ is to 160?

$\frac{\text{number}}{100} = \frac{\text{part}}{\text{whole}}$ Set up a ____.

$\frac{n}{100} = \frac{\_\_\_\_}{160}$ Substitute.

$n \cdot 160 = 100 \cdot$ ____ Find the cross ____.

$160n =$ ____

$n = \frac{\_\_\_\_}{160}$ Solve for n.

$n =$ ____

$\frac{\_\_\_\_}{100} = \frac{\_\_\_\_}{160}$ The proportion is reasonable.

So ____ % of Eddie's weight is bone.

Example 2

A. After a drought, a reservoir had only $66\frac{2}{3}\%$ of the average amount of water. If the average amount of water is 57,000,000 gallons, how much water was in the reservoir after the drought?

Choose a method: Set up an equation.

Think: What number is $66\frac{2}{3}\%$ of 57,000,000?

$w = 66\frac{2}{3}\% \cdot$ Set up an equation.

$w = \quad \cdot 57{,}000{,}000$ $66\frac{2}{3}\%$ is equivalent to

$w = \frac{114{,}000{,}000}{3} =$

The reservoir contained gallons of water after the drought.

Try This

1. What percent of 110 is 11?

2. After a drought, a river had only $50\frac{2}{3}\%$ of the average amount of water flow. If the average amount of water flow is 60,000,000 gallons per day, how much water was flowing in the river after the drought?

LESSON 8-3 Finding a Number When the Percent is Known *pp. 410–411*

Additional Examples

Example 1

60 is 12% of what number?

Set up an equation to find the number.

$60 = 12\% \cdot$ ____ Set up an equation.

$60 = 0.12$ ____ $12\% = \frac{\quad}{100}$

____ = ____ n Divide both sides by ____.

____ $= n$ 60 is 12% of ____.

Example 2

Anna earned 85% on a test by answering 17 questions correctly. If each question was worth the same amount, how many questions were on the test?

Choose a method: Set up a proportion to find the number.

Think: 85 is to ____ as 17 is to what number?

____ $= \frac{17}{n}$ Set up a ____.

$85 \cdot n =$ ____ $\cdot 17$ Find the cross ____.

$85n =$ ____

$n = \frac{\quad}{85}$ Solve for n.

$n =$ ____

There were ____ questions on the test.

Example 3

A. A fisherman caught a lobster that weighed 11.5 lb. This was 70% of the weight of the largest lobster that fisherman had ever caught. What was the weight, to the nearest tenth of a pound, of the largest lobster the fisherman had ever caught?

Choose a method: Set up an equation.

Think: 11.5 is % of what number?

$11.5 = \quad \% \cdot n$ Set up an equation.

$11.5 = 0. \quad \cdot n$ $70\% = 0.$

$= n$ Solve for n.

$\approx n$

The largest lobster the fisherman had ever caught was about lb.

Try This

1. 75 is 25% of what number?

2. Tom earned 80% on a test by answering 20 questions correctly. If each question was worth the same amount, how many questions were on the test?

3. When Bart was 12, he was approximately 85% of the weight he is now. If Bart was 120 lb, how heavy is he now, to the nearest tenth of a pound?

LESSON 8-4

Percent Increase and Decrease

pp. 416–417

Vocabulary

percent change (p. 416) ______________________

percent increase (p. 416) ______________________

percent decrease (p. 416) ______________________

Additional Examples

Example 1

Find the percent increase or decrease from 16 to 12.

This is percent decrease.

$16 - 12 =$ ☐ First find the amount of ☐ .

Think: What percent is ☐ of 16?

$\frac{\text{amount of decrease}}{\text{original amount}} \rightarrow \frac{\square}{16}$ Set up the ratio.

$\frac{4}{16} =$ ☐ Find the decimal form.

$=$ ☐ % Write as a percent.

From 16 to 12 is a ☐ % ☐ .

Example 3

A. Sarah bought a DVD player originally priced at $450 that was on sale for 20% off. What was the sale price?

$450 • 20% First find % of $450.

$450 • 0. = $ 20% = 0.

The amount of decrease is $.

Think: The reduced price is $ less than $450.

$450 − $ = $ Subtract the amount of decrease.

The sale price of the DVD player was $.

Try This

1. Find the percent increase or decrease from 15 to 20.

2. Lily bought a dog house originally priced at $750 that was on sale for 10% off. What was the sale price?

LESSON **8-5**

Estimating with Percents

pp. 420–421

Vocabulary

estimate (p. 420) ______________________________

____

compatible numbers (p. 420) ______________________________

____

Additional Examples

Example 1

Estimate.

A. 21% of 66

Instead of computing the exact answer of 21% • 66, estimate.

$21\% = \underline{\qquad} \approx \frac{20}{100}$ Use ______ numbers, 20 and 100.

$\approx \underline{\qquad}$ Simplify.

$66 \approx 65$ Use ______ numbers, 65 and 5.

$\frac{1}{5} \cdot 65 = \underline{\qquad}$ Use mental math: 65 ÷ 5.

So 21% of 66 is about ______.

Example 2

PROBLEM SOLVING APPLICATION

The diameter of the Moon is about 2160 miles. If the diameter of the Moon is about 27% of the diameter of Earth, what is the approximate diameter of Earth?

1. Understand the Problem

The answer is the approximate   of Earth.
List the important information:

- The diameter of the Moon is about   miles.
- The diameter of the Moon is about   % of the diameter of the Earth.

Let e represent the diameter of the Earth.

Diameter of the Moon	$\approx$	27%	$\cdot$	Diameter of Earth
2160	$\approx$	27%	$\cdot$	e

2. Make a Plan

Think: The numbers 2160 and 27% are difficult to work with.
Use compatible numbers: 2160 is close to 2000; 27% is close to 25%.

25% =   =   Find an equivalent ratio for 25%.

3. Solve

Think: 2000 is   of what number?

$4 \cdot 2000 = e$

$= e$

The diameter of Earth is approximately   miles.

4. Look Back

25% of 8000 mi is $\frac{8000}{4}$, or 2000 mi. This is the approximate diameter of the moon.

Try This

2. Problem Solving Application

The length of a football field is 83% of the length of a soccer field. If a football field is 100 yards, what is the approximate length of a soccer field?

1. Understand the Problem

The answer is the approximate ________ of a soccer field.

List the important information:

- The length of a football field is about ________ yards.
- The length of a football field is ________ % of a soccer field.

Let l represent the length of a soccer field.

Length of football field	$\approx$	83%	•	Length of soccer field
100	$\approx$	83%	•	l

2. Make a Plan

Think: The number 83% is difficult to work with.
Use compatible numbers: 83% is close to 80%.

80% = ________ = ________ Find an equivalent ratio for 80%.

3. Solve

Think: 100 is $\frac{4}{5}$ of what number?

$\frac{5}{4} \cdot 100 = l$

________ $= l$

____

4. Look Back

80% of 125 yards is $\frac{4}{5} \cdot 125$, or 100 yards. This is the length of a football field.

LESSON **8-6**

Applications of Percents

pp. 424–425

Vocabulary

commission (p. 424)

commission rate (p. 424)

sales tax (p. 424)

withholding tax (p. 425)

Additional Examples

Example 1

A real-estate agent is paid a monthly salary of \$900 plus commission. Last month he sold one condominium for \$65,000, earning a 4% commission on the sale. How much was his commission? What was his total pay last month?

First find his commission.

4% • \$	$= c$	commission rate • sales = commission
0.04 •	$= c$	Change the percent to a decimal.
	$= c$	Solve for c.

He earned a commission of \$ on the sale.

Now find his total pay for last month.

\$	+ \$	= \$	commission + salary = total pay

His total pay for last month was \$.

Example 2

If the sales tax rate is 6.75%, how much tax would Adrian pay if he bought two CDs at $16.99 each and one DVD for $36.29?

CD: 2 at $16.99 ⟶ $ ______

DVD: 1 at $36.29 ⟶ $ ______

$ ______ Total price

$0.0675 \cdot 70.27 = 4.743225$ Convert tax rate to a decimal and multiply by the total price.

Adrian would pay $ ______ in sales tax.

Example 3

Anna earns $1500 monthly. Of that, $114.75 is withheld for Social Security and Medicare. What percent of Anna's earnings are withheld for Social Security and Medicare?

Think: What percent of $ ______ is $114.75?

Solve by proportion:

$$\frac{n}{\square} = \frac{114.75}{\square}$$

$n \cdot \square = \square \cdot 114.75$ Find the cross products.

$\square = \square$ Divide both sides by ______.

$n = \square$

$n = \square$

______ % of Anna's earnings is withheld for Social Security and Medicare.

Example 4

A furniture sales associate earned $960 in commission in May. If his commission is 12% of sales, how much were his sales in May?

Think: $960 is   % of what number?

Solve by equation:

$960 = 0.12 \cdot s$ Let s = total sales.

$\frac{960}{\quad} = s$ Divide each side by   .

The associate's sales in May were $.

Try This

1. **A car sales agent is paid a monthly salary of $700 plus commission. Last month she sold one sports car for $50,000, earning a 5% commission on the sale. How much was her commission? What was her total pay last month?**

2. **Amy rents a hotel room for $45 per night. She rents for two nights and pays a sales tax of 13%. How much tax did she pay?**

3. **BJ earns $2500 monthly. Of that, $500 is withheld for income tax. What percent of BJ's earnings are withheld for income tax?**

4. **A sales associate earned $770 in commission in May. If his commission is 7% of sales, how much were his sales in May?**

LESSON 8-7

More Applications of Percents

pp. 428–429

Vocabulary

interest (p. 428) ______

simple interest (p. 428) ______

principal (p. 428) ______

rate of interest (p. 428) ______

Additional Examples

Example 1

To buy a car, Jessica borrowed $15,000 for 3 years at an annual simple interest rate of 9%. How much interest will she pay if she pays the entire loan off at the end of the third year? What is the total amount that she will repay?

First, find the interest she will pay.

$I = P \cdot r \cdot t$ Use the formula.

$I = \square \cdot \square \cdot \square$ Substitute. Use 0.09 for 9%.

$I = \square$ Solve for I.

Jessica will pay $ ______ in interest.

You can find the total amount A to be repaid on a loan by adding the principal P to the interest I.

$P + I = A$ principal + interest = amount

$\square + \square = A$ Substitute.

$\square = A$ Solve for A.

Jessica will repay a total of $ ______ on her loan.

Example 4

Mr. Johnson borrowed $8000 for 4 years to make home improvements. If he repaid a total of $10,320, at what interest rate did he borrow the money?

$P + I = A$ Use the formula.

$+ I =$

$I = 10{,}320 - 8000 =$ Find the amount of interest.

He paid $ in interest. Use the amount of interest to find the interest rate.

$I = P \cdot r \cdot t$ Use the formula.

$= \quad \cdot r \cdot$ Substitute.

$= \quad \cdot r$ Multiply.

$\frac{2320}{\quad} = r$ Divide both sides by 32,000.

$= r$

Mr. Johnson borrowed the money at an annual rate of %, or %.

Try This

1. **To buy a laptop computer, Elaine borrowed $2,000 for 3 years at an annual simple interest rate of 5%. How much interest will she pay if she pays the entire loan off at the end of the third year? What is the total amount that she will repay?**

Chapter 8
Tetra Terms

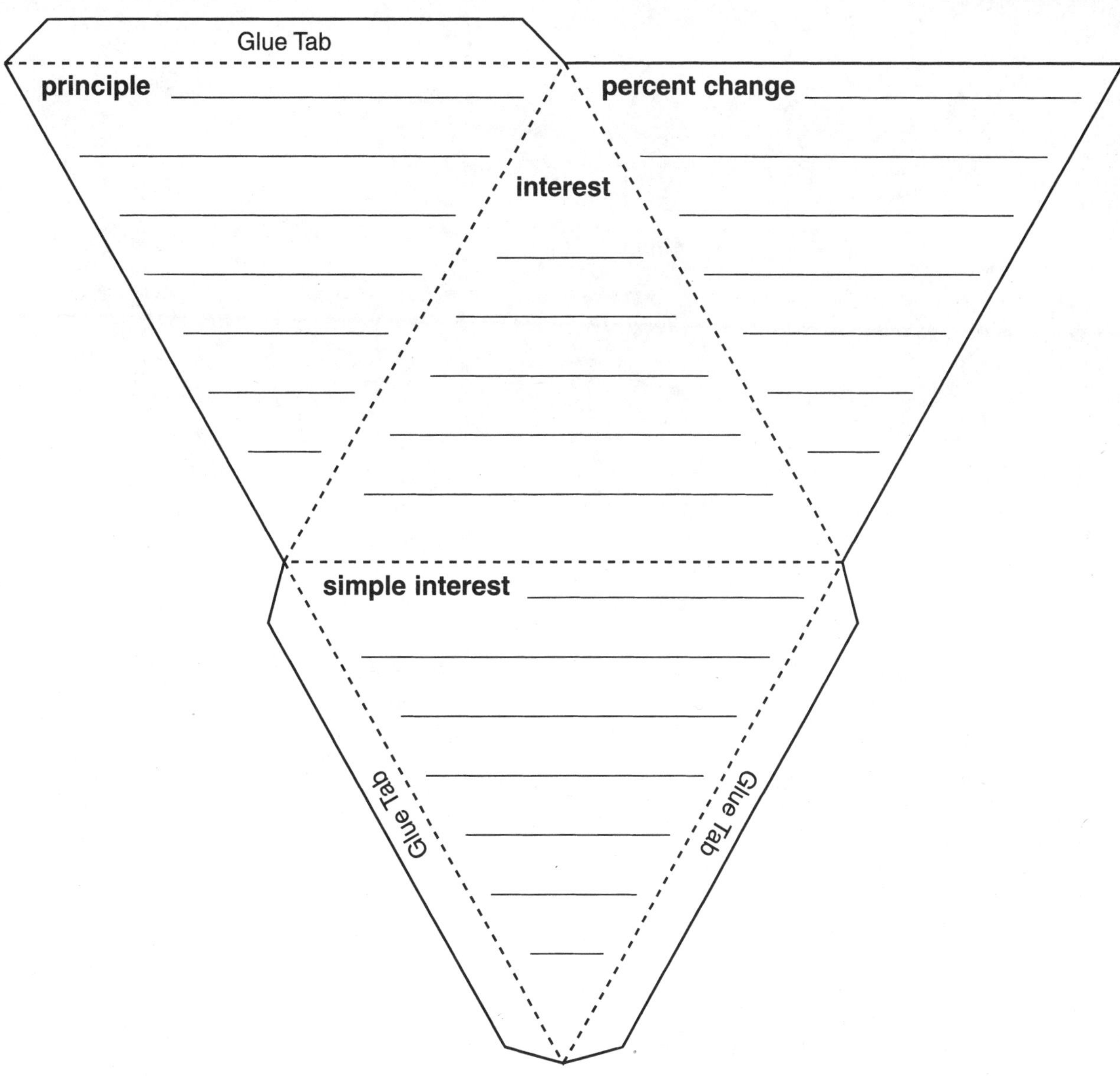

Directions

1. Write the definition of each term on the net of the tetrahedron.
2. Cut out the net.
3. Fold along all dotted lines, and place glue tabs to the inside of the tetrahedron.
4. Join the common edges, and tape or glue the tabs in place.

LESSON **9-1** **Probability**
pp. 446–448

Vocabulary

experiment (p. 446)

trial (p. 446)

outcome (p. 446)

sample space (p. 446)

event (p. 446)

probability (p. 446)

impossible (p. 446)

certain (p. 446)

Additional Examples

Example 1

Give the probability for the outcome.

The basketball team has a 70% chance of winning.

The probability of winning is P(win) = ____ % = ____. The probabilities must add to ____, so the probability of not winning is P(lose) = ____ − ____ = ____, or ____ %.

Example 3

PROBLEM SOLVING APPLICATION

Six students are in a race. Ken's probability of winning is 0.2. Lee is twice as likely to win as Ken. Roy is $\frac{1}{4}$ as likely to win as Lee. Tracy, James, and Kadeem all have the same chance of winning. Create a table of probabilities for the sample space.

1. Understand the Problem

The answer will be a table of probabilities. Each probability will be a number from 0 to 1. The probabilities of all outcomes add to 1. List the important information:

- $P(\text{Ken}) =$
- $P(\text{Lee}) = 2 \cdot P(\text{Ken}) = 2 \cdot 0.2 =$
- $P(\text{Roy}) = \frac{1}{4} \cdot P(\text{Lee}) = \frac{1}{4} \cdot \quad =$
- $P(\text{Tracy}) = P(\text{James}) = P(\text{Kadeem})$

2. Make a Plan

You know the probabilities add to 1, so use the strategy write an equation. Let p represent the probability for Tracy, James, and Kadeem.

$$P(\text{Ken}) + P(\text{Lee}) + P(\text{Roy}) + P(\text{Tracy}) + P(\text{James}) + P(\text{Kadeem}) = 1$$

$$0.2 + 0.4 + 0.1 + p + p + p = 1$$

$$0.7 + 3p = 1$$

3. Solve

$0.7 + 3p = 1$

$-\square \quad -\square$ Subtract $\square$ from both sides.

$3p = \square$

$\frac{3p}{\square} = \frac{0.3}{\square}$ Divide both sides by $\square$.

$p = \square$

Outcome	Ken	Lee	Roy	Tracy	James	Kadeem
Probability						

4. Look Back

Check that the probabilities add to 1.

$0.2 + 0.4 + 0.1 + 0.1 + 0.1 + 0.1 = 1$ ✓

Try This

3. Problem Solving Application

Four students are in the Spelling Bee. Fred's probability of winning is 0.6. Willa's chances are one-third of Fred's. Betty's and Barrie's chances are the same. Create a table of probabilities for the sample space.

Outcome	Fred	Willa	Betty	Barrie
Probability				

LESSON 9-2

Experimental Probability

pp. 451–452

Vocabulary

experimental probability (p. 451)

Additional Examples

Example 1

A. The table shows the results of 500 spins of a spinner. Estimate the probability of the spinner landing on 2.

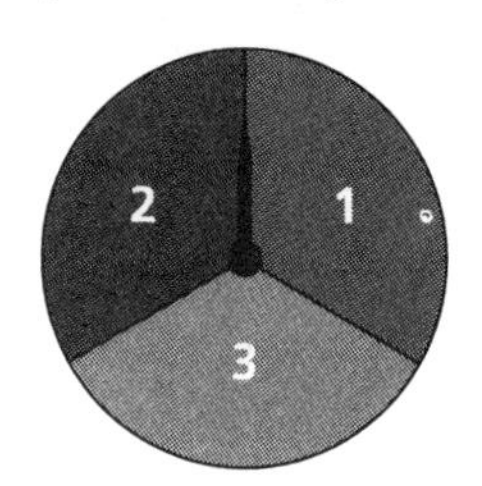

Outcome	1	2	3
Spins	151	186	163

$$\text{probability} \approx \frac{\text{number of spins that landed on \_\_\_\_\_}}{\text{total number of spins}} = \_\_\_\_\_$$

The probability of landing on 2 is about ______, or ______%.

B. A customs officer at the New York–Canada border noticed that of the 60 cars that he saw, 28 had New York license plates, 21 had Canadian license plates, and 11 had other license plates. Estimate the probability that a car will have Canadian license plates.

Outcome	New York	Canadian	Other
Observations	28	21	11

$$\text{probability} \approx \frac{\text{number of \_\_\_\_\_ license plates}}{\text{total number of license plates}} = \_\_\_\_\_ = \_\_\_\_\_$$

The probability that a car will have Canadian license plates is about ______,

or ______%.

Example 2

Team	Wins	Games
Huskies	79	138
Cougars	85	150
Knights	90	146

Use the table to compare the probability that the Huskies will win their next game with the probability that the Knights will win their next game.

probability $\approx \frac{\text{number of wins}}{\text{number of games}}$

probability for a Huskies win $\approx \frac{\square}{\square} \approx \square$

probability for a Knights win $\approx \frac{\square}{\square} \approx \square$

The Knights are ______ likely to win their next game than the Huskies.

Try This

1. **Jeff tosses a quarter 1000 times and finds that it lands heads 523 times. What is the probability that the next toss will land heads? Tails?**

2. **Use the above table to compare the probability that the Huskies will win their next game with the probability that the Cougars will win their next game.**

LESSON 9-3

Use a Simulation

pp. 456–457

Vocabulary

simulation (p. 456)

random numbers (p. 456)

Additional Examples

Example 1

PROBLEM SOLVING APPLICATION

A dart player hits the bull's-eye 25% of the times that he throws a dart. Estimate the probability that he will make at least 2 bull's-eyes out of his next 5 throws.

1. Understand the Problem

The answer will be the that he will make at least 2 bull's-eyes out of his next 5 throws. List the important information:

- The probability that the player will hit the bull's-eye is .

2. Make a Plan

Use a to model the situation. Use digits grouped in pairs. The numbers 01–25 represent a bull's-eye, and the numbers 26–00 represent an unsuccessful attempt. Each group of 10 digits represent one trial.

87244	11632	85815	61766	19579	28186	18533	42633
74681	65633	54238	32848	87649	85976	13355	46498
53736	21616	86318	77291	24794	31119	48193	44869
86585	27919	65264	93557	94425	13325	16635	28584
18394	73266	67899	38783	94228	23426	76679	41256
39917	16373	59733	18588	22545	61378	33563	65161
96916	46278	78210	13906	82794	01136	60848	98713

3. Solve
Starting on the third row of the table and using 10 digits for each trial yiel the following data:

53	73	62				bull's eyes
86	31	87	72	91		bull's eyes
	79	43				bull's eyes
48		34	48	69		bull's eyes
86	58	52	79			bull's eyes
65	26	49	35	57		bull's eyes
94	42	51	33			bull's eyes
	63	52	85	84		bull's eyes
	39	47	32	66		bull's eyes
67	89	93	87	83		bull's eyes

Out of the 10 trials, ______ trials represented two or more bull's-eyes.

Based on this simulation, the probability of making at least 2 bull's-eyes out of his next 5 throws is about ______, or ______ %.

4. Look Back
Hitting the bull's-eye at a rate of 20% means the player hits about ______ bull's-eyes out of every 100 throws. This ratio is equivalent to 2 out of 10 throws, so he should make at least 2 bull's-eyes most of the time. The answer is reasonable.

LESSON 9-4 Theoretical Probability
pp. 462–464

Vocabulary

theoretical probability (p. 462)

equally likely (p. 462)

fair (p. 462)

mutually exclusive (p. 464)

Additional Examples

Example 1

An experiment consists of spinning this spinner once.

A. What is the probability of spinning a 4?

The spinner is , so all 5 outcomes are equally likely. The probability of spinning a 4 is $P(4) =$.

B. What is the probability of spinning an even number?

There are outcomes in the event of spinning an even number: and .

$P(\text{spinning an even number}) = \frac{\text{number of possible } \qquad \text{ numbers}}{5} =$

Example 2

An experiment consists of rolling one fair die and flipping a coin.

A. Show a sample space that has all outcomes equally likely.

The outcome of rolling a 5 and flipping heads can be written as the ordered pair (5, H). There are ____ possible outcomes in the sample space.

B. What is the probability of getting tails?

There are ____ outcomes in the event "flipping tails":

P(tails) = ____ = ____

C. What is the probability of getting an even number and heads?

There are ____ outcomes in the event "getting an even number and heads":

P(even number and heads) = ____ = ____

D. What is the probability of getting a prime number?

There are ____ outcomes in the event "getting a prime number":

P(prime number) = ____ = ____

Example 3

Suppose you are playing a game in which you roll two fair dice. If you roll a total of five you will win. If you roll a total of two, you will lose. If you roll anything else, the game continues. What is the probability that the game will end on your next roll?

It is impossible to roll a total of 5 and a total of 2 at the same time, so the events are ______ exclusive. Add the probabilities to find the probability of the game ending on your next roll.

The event "total = 5" consists of ___ outcomes,

______, so P(total = 5) = ___.

The event "total = 2" consists of ___ outcome, ___,

so P(total = 2) = ___.

P(game ends) = P(total = 5) + P(total = 2)

= ___ + ___ = ___

The probability that the game will end is ___, or about ___%.

Try This

1. An experiment consists of spinning this spinner once.

What is the probability of spinning an odd number?

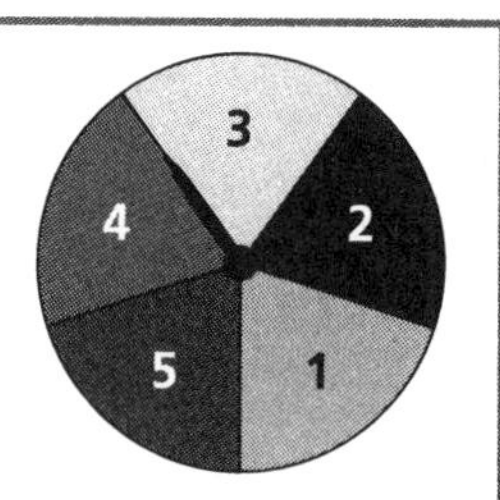

2. An experiment consists of flipping two coins.

What is the probability of getting one head and one tail?

LESSON **9-5**

The Fundamental Counting Principle

pp. 467–468

Know-It Notes

Vocabulary

Fundamental Counting Principle (p. 467) ______

tree diagram (p. 468) ______

Additional Examples

Example 1

License plates are being produced that have a single letter followed by three digits. All license plates are equally likely.

A. Find the number of possible license plates.

Use the ______ Counting Principle.

letter	first digit	second digit	third digit
■	■	■	■
___ choices	___ choices	___ choices	___ choices

___ • ___ • ___ • ___ = ___

The number of possible 1-letter, 3-digit license plates is ______.

B. Find the probability that a license plate has the letter Q.

$P(Q \blacksquare \blacksquare \blacksquare) = \frac{1 \cdot 10 \cdot 10 \cdot 10}{\_\_\_} = \frac{1}{26} \approx$ ______

Example 2

You have a photo that you want to mat and frame. You can choose from a blue, purple, red, or green mat and a metal or wood frame. Describe all of the ways you could frame this photo with one mat and one frame.

You can find all of the possible outcomes by making a tree ______.

There should be ___ · ___ = ___ different ways to frame the photo.

Each "branch" of the tree diagram represents a different way to frame the photo. The ways shown in the branches could be written as

.

Try This

1. Social Security numbers contain 9 digits. All social security numbers are equally likely.

Find the probability that the Social Security number contains a 7.

LESSON 9-6

Permutations and Combinations

pp. 471–473

Know-It Notes

Vocabulary

factorial (p. 471)

permutation (p. 471)

combination (p. 472)

Additional Examples

Example 1

Evaluate each expression.

A. 8!

☐ · ☐ · ☐ · ☐ · ☐ · ☐ · ☐ · ☐ = ☐

B. $\frac{8!}{6!}$

Write out each ☐ and simplify.

☐ · ☐ = ☐ Multiply remaining factors.

Example 2

Jim has 6 different books.

A. Find the number of orders in which the 6 books can be arranged on a shelf.

The number of books is .

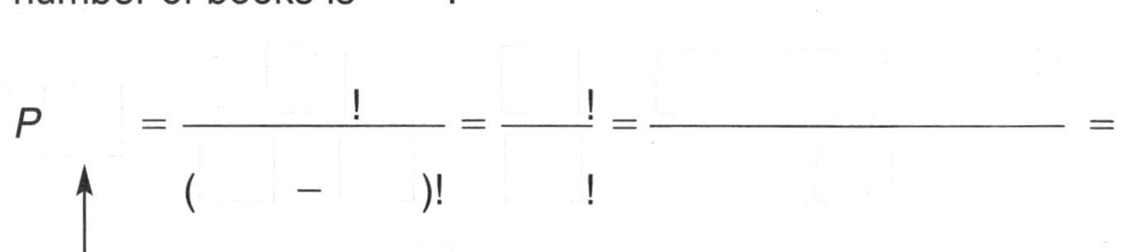

The books are arranged at a time.

There are permutations. This means there are orders in which the 6 books can be arranged on the shelf.

B. If the shelf has room for only 3 of the books, find the number of ways 3 of the 6 books can be arranged.

The number of books is .

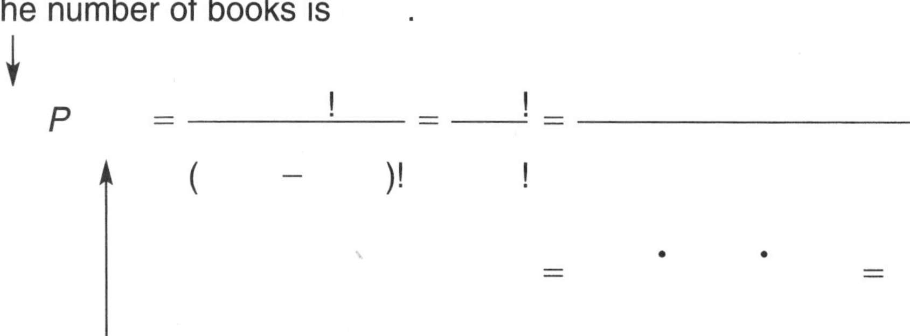

The books are arranged at a time.

There are permutations. This means that 3 of the 6 books can be arranged in ways.

Example 3

Mary wants to join a book club that offers a choice of 10 new books each month.

A. If Mary wants to buy 2 books, find the number of different pairs she can buy.

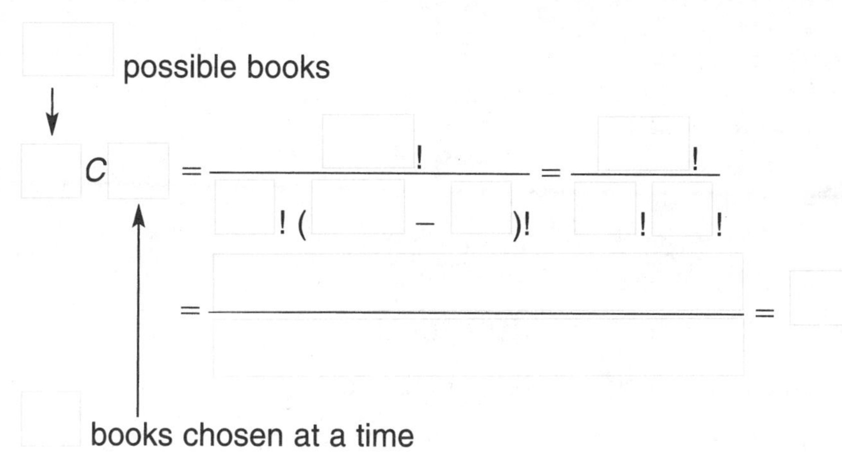

There are ______ combinations. This means that Mary can buy ______ different pairs of books.

Try This

1. Evaluate the expression.

$$\frac{9!}{(8-2)!}$$

2. There are 7 soup cans in the pantry. If the shelf has only enough room for 4 cans, find the number of ways 4 of the 7 cans can be arranged.

LESSON **9-7**

Independent and Dependent Events

pp. 477–479

Vocabulary

independent events (p. 477)

dependent events (p. 477)

Additional Examples

Example 1

Determine if the events are dependent or independent.

A. getting tails on a coin toss and rolling a 6 on a number cube

Tossing a coin does not affect rolling a number cube, so the two events are

.

B. getting 2 red gumballs out of a gumball machine

After getting one red gumball out of a gumball machine, the chances for getting the second red gumball have changed, so the two events are

.

Example 2

Three separate boxes each have one blue marble and one green marble. One marble is chosen from each box.

A. What is the probability of choosing a blue marble from each box?

The outcome of each choice does not affect the outcome of the other choices, so the choices are .

In each box, P(blue) = .

P(blue, blue, blue) = · · = = Multiply.

B. What is the probability of choosing a blue marble, then a green marble, and then a blue marble?

In each box, $P(\text{blue}) = \frac{1}{2}$.

In each box, $P(\text{green}) = \frac{1}{2}$.

$P(\text{blue, green, blue}) = \square \cdot \square \cdot \square = \square = \square$ Multiply.

C. What is the probability of choosing at least one blue marble?
Think: $P(\text{at least one blue}) + P(\text{not blue, not blue, not blue}) = 1$.

In each box, $P(\text{not blue}) = \square$.

$P(\text{not blue, not blue, not blue}) = \square \cdot \square \cdot \square$

$= \square = \square$ Multiply.

Subtract from 1 to find the probability of choosing at least one blue marble.

$1 - 0.125 = \square$

Example 3

The letters in the word *dependent* are placed in a box.

If two letters are chosen at random, what is the probability that they will both be consonants?

$P(\text{first consonant}) = \square = \square$

If the first letter chosen was a consonant, now there would be 5 consonants and a total of 8 letters left in the box. Find the probability that the second letter chosen is a consonant.

$P(\text{second consonant}) = \square$

$\square \cdot \square = \square$ Multiply.

The probability of choosing two letters that are both consonants is $\square$.

Try This

1. Determine if the events are dependent or independent.

rolling a 6 two times in a row with the same number cube

2. Two boxes each contain 4 marbles: red, blue, green, and black. One marble is chosen from each box.

What is the probability of choosing a blue marble and then a red marble?

3. The letters in the phrase *I Love Math* are placed in a box.

If two letters are chosen at random, what is the probability that they will both be consonants?

LESSON **9-8**

Odds

pp. 482–483

Vocabulary

odds in favor (p. 482) ______________________________

odds against (p. 482) ______________________________

Additional Examples

Example 1

In a club raffle, 1,000 tickets were sold, and there were 25 winners.

A. Estimate the odds in favor of winning this raffle.

The number of ________ outcomes is 25, and the number of ________ outcomes is $1000 - 25 = 975$. The odds in favor of winning this raffle are about ____ to ____, or ____ to ____.

B. Estimate the odds against winning this raffle.

The odds in ________ of winning this raffle are 1 to 39, so the odds against winning this raffle are about ____ to ____.

Example 2

A. If the odds in favor of winning a CD player in a school raffle are 1:49, what is the probability of winning a CD player?

$P(\text{CD player}) = \frac{1}{1 + 49} =$ ____

On average there is 1 win for every ____ losses, so someone wins 1 out of every ____ times.

B. If the odds against winning the grand prize are 11,999:1, what is the probability of winning the grand prize?

If the odds winning the grand prize are 11,999:1, then

the odds in favor of winning the grand prize are .

$P(\text{grand prize}) = \frac{1}{1 + 11{,}999} = \frac{1}{12{,}000} \approx$

Example 3

A. The probability of winning a free dinner is $\frac{1}{20}$. What are the odds in favor of winning a free dinner?

On average, 1 out of every people wins, and the other 19 people

lose. The odds in favor of winning the meal are 1:(20 − 1), or .

B. The probability of winning a door prize is $\frac{1}{10}$. What are the odds against winning a door prize?

On average, out of every 10 people wins, and the other 9 people lose.

The odds against the door prize are (10 − 1):1, or .

Try This

1. Of the 1750 customers at an arts and crafts show, 25 will win door prizes. Estimate the odds in favor winning a door prize.

2. If the odds in favor of winning a bicycle in a raffle are 1:75, what is the probability of winning a bicycle?

Chapter 9

Vocabulary Chain

EXAMPLE		TERM
		dependent events
EXAMPLE		TERM
		experimental probability
EXAMPLE		TERM
		independent events
EXAMPLE		TERM
		permutation
EXAMPLE		TERM
		theoretical probability

Directions

1. Write an example and an explanation in words, numbers, and algebra for each term.
2. Cut out each vocabulary strip and fold in thirds.
3. Punch a hole in the corner of each folded vocabulary strip, and string them together to create your vocabulary chain.

Chapter 9
Vocabulary Chain

WORDS	NUMBERS	ALGEBRA
WORDS	NUMBERS	ALGEBRA
WORDS	NUMBERS	ALGEBRA
WORDS	NUMBERS	ALGEBRA
WORDS	NUMBERS	ALGEBRA

LESSON 10-1

Solving Two-Step Equations

pp. 498–499

Additional Examples

Example 1

PROBLEM SOLVING APPLICATION

The mechanic's bill to repair Mr. Wong's car was $650. The mechanic charges $45 an hour for labor, and the parts that were used cost $443. How many hours did the mechanic work on the car?

1. Understand the Problem

List the important information:

The answer is the number of ______ the mechanic worked on the car.

- The parts cost $______.
- The labor cost $______ per hour.
- The total bill was $______.

Let h represent the hours the mechanic worked.

Total bill = Parts + Labor

$\_\_\_\_ = \_\_\_\_ + \_\_\_\_ h$

2. Make a Plan

Think: First the variable is multiplied by ______, and then ______ is added to the result. Work backward to solve the equation. Undo the operations in reverse order: First subtract ______ from both sides of the equation, and then divide both sides of the new equation by ______.

3. Solve

= + h

− − Subtract to undo the addition.

= h

= Divide to undo multiplication.

= h

The mechanic worked for hours on Mr. Wong's car.

4. Look Back

If the mechanic worked hours, the labor would be $45(4.6) = $207.

The sum of the parts and the labor would be $ + $ =

$.

Example 2

Solve.

$\frac{n}{3} + 7 = 22$

Think: First the variable is by 3, and then 7 is .

To isolate the variable, subtract , and then multiply by .

$\frac{n}{3} + 7 = \quad 22$

− − Subtract to undo addition.

$\frac{n}{3} \quad =$

$\cdot \frac{n}{3} = \quad \cdot 15$ Multiply to undo division.

$n =$ ☐

Check

$\frac{n}{3} + 7 = 22$

$\frac{\square}{3} + 7 \stackrel{?}{=} 22$ Substitute ☐ into the original equation.

$\square + 7 \stackrel{?}{=} 22$ ✓

Try This

1. Problem Solving Application

The mechanic's bill to repair your car was $850. The mechanic charges $35 an hour for labor, and the parts that were used cost $275. How many hours did the mechanic work on your car?

LESSON 10-2 Solving Multistep Equations

pp. 502–503

Additional Examples

Example 1

Solve.

$8x + 6 + 3x - 2 = 37$

$x + \quad = 37$ Combine like terms.

$-4 \quad -4$ Subtract to undo .

$x = 33$

$\frac{11x}{11} = \frac{33}{11}$ to undo multiplication.

$x =$

Check

$8x + 6 + 3x - 2 = 37$

$8(\quad) + 6 + 3(\quad) - 2 \overset{?}{=} 37$ Substitute for x.

$+ 6 + \quad - 2 \overset{?}{=} 37$

$\overset{?}{=} 37$

Example 2

Solve.

$\frac{5n}{4} + \frac{7}{4} = -\frac{3}{4}$

Multiply both sides by 4 to clear fractions, and then solve.

$\square\left(\frac{5n}{4} + \frac{7}{4}\right) = \square\left(\frac{-3}{4}\right)$

$4\left(\frac{5n}{4}\right) + 4\left(\frac{7}{4}\right) = 4\left(\frac{-3}{4}\right)$ ______ Property.

$\square n + \square = \square$

$\underline{-7} \quad \underline{-7}$ ______ to undo addition.

$\square n = \square$

$\frac{5n}{5} = \frac{-10}{5}$ Divide to undo ______.

$n = \square$

Try This

1. Solve.

$9x + 5 + 4x - 2 = 42$

2. Solve.

$\frac{5x}{9} + \frac{x}{3} - \frac{13}{9} = \frac{1}{3}$

LESSON **10-3**

Solving Equations with Variables on Both Sides *pp. 507–509*

Additional Examples

Example 1

Solve.

A. $4x + 6 = x$

$4x + 6 = \quad x$

$-$ ______ $\quad -$ ______ Subtract from both sides.

$6 = \qquad x$

$\dfrac{6}{\quad} = \dfrac{-3x}{\quad}$ Divide both sides by .

$= x$

B. $9b - 6 = 5b + 18$

$9b - 6 = \quad 5b + 18$

$-$ ______ $\quad -$ ______ Subtract from both sides.

$- 6 = \quad 18$

$+$ ______ $\quad +$ ______ Add to both sides.

$4b =$

$\dfrac{4b}{\quad} = \dfrac{24}{\quad}$ Divide both sides by .

$b =$

Example 2

Solve.

$10z - 15 - 4z = 8 - 2z - 15$

$10z - 15 - 4z = 8 - 2z - 15$

$\square - 15 = -2z - \square$ Combine like terms.

$+\square \qquad +\square$ Add $\square$ to both sides.

$\square - 15 = \qquad -7$

$+\square \qquad +\square$ Add $\square$ to both sides.

$8z = \square$

$\frac{8z}{\square} = \frac{8}{\square}$ Divide both sides by $\square$.

$z = \square$

Try This

1. Solve.

$3b - 2 = 2b + 12$

2. Solve.

$\frac{y}{4} + \frac{5y}{6} + \frac{3}{4} = y - \frac{6}{8}$

Solving Multistep Inequalities

pp. 514–516

Additional Examples

Example 1

Solve and graph.

A. $4x + 1 > 13$

$4x + 1 > \quad 13$

$\underline{-\quad\quad} \quad \underline{-\quad\quad}$ Subtract from both sides.

$4x \quad\quad >$

$\frac{4x}{\quad} > \frac{12}{\quad}$ Divide both sides by .

$x >$

C. $-9x + 7 \geq 25$

$-9x + 7 \geq \quad 25$

$\underline{-\quad\quad} \quad \underline{-\quad\quad}$ Subtract from both sides.

$-9x \quad\quad \geq$

$\frac{-9x}{\quad} \quad \frac{18}{\quad}$ Divide both sides by ; change to .

x

−6 −5 −4 −3 −2 −1 0

Know-It Notes

Example 2

Solve and graph.

A. $10x + 21 - 4x < -15$

$10x + 21 - 4x < -15$

$\square + 21 < -15$ Combine like terms.

$- \square \quad - \square$ Subtract $\square$ from both sides.

$6x < \square$

$\frac{6x}{\square} < \frac{-36}{\square}$ Divide both sides by $\square$.

$x < \square$

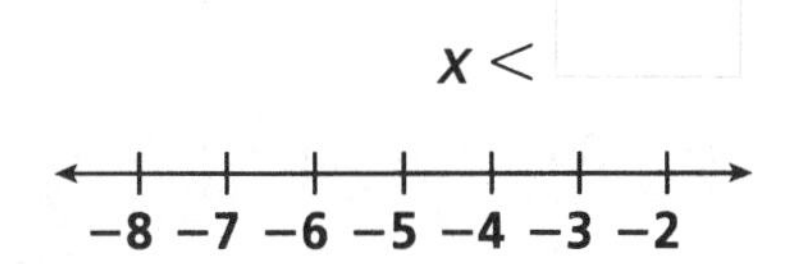

Try This

1. Solve and graph.

$-4x + 2 \geq 18$

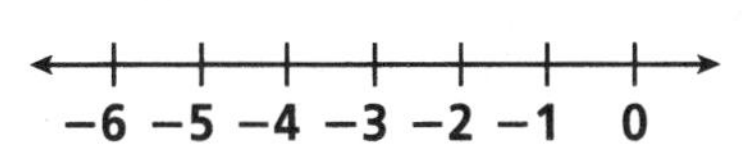

2. Solve and graph.

$\frac{3x}{5} + \frac{1}{4} \geq \frac{5}{10}$

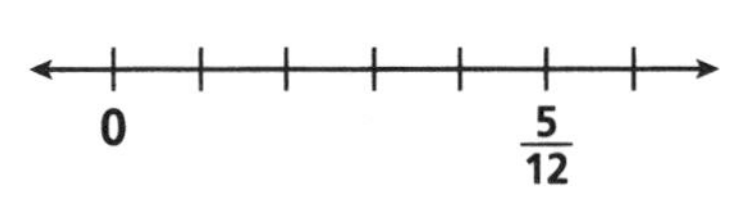

Solving for a Variable

pp. 519–520

Additional Examples

Example 1

Solve for the indicated variable.

A. Solve $a - b + 1 = c$ for a.

$a - b \quad + 1 = c$

$+ \quad - \qquad + \quad -$ Add and subtract from both sides.

$a \qquad =$ Isolate .

B. Solve $a - b + 1 = c$ for b.

$a - b + 1 = c$

$- \quad - \qquad - \quad -$ Subtract and from both sides.

$-b \quad = c - a - 1$ Isolate b.

$\cdot (-b) = \qquad \cdot (c - a - 1)$ Multiply both sides by .

$b =$ Isolate b.

Example 2

Solve the formula for the area of a trapezoid for *h*. Assume all values are positive.

$A = \frac{1}{2}h(b_1 + b_2)$ Write the formula.

$\square \cdot A = \square \cdot \frac{1}{2}h(b_1 + b_2)$ Multiply both sides by $\square$.

$2A = h(b_1 + b_2)$

$\frac{2A}{\square} = \frac{h(b_1 + b_2)}{\square}$ Divide both sides by $\square$.

$\square = h$ Isolate $\square$.

Try This

1. Solve for the indicated variable.

Solve $p - w + 4 = f$ for w.

2. Solve for the indicated variable. Assume all values are positive.

Solve $s = 180(n - 2)$ for n.

3. Solve for y and graph $4x + 3y = 12$.

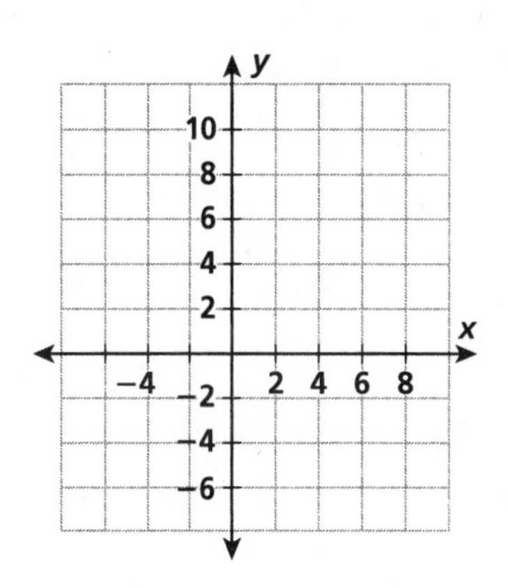

System of Equations

pp. 523–525

Vocabulary

system of equations (p. 523)

solution of a system of equations (p. 523)

Additional Examples

Example 1

Determine if each ordered pair is a solution of the system of equations below.

$$5x + y = 7$$
$$x - 3y = 11$$

A. (1, 2)

$5x + y = 7$ $\qquad$ $x - 3y = 11$

$5(\quad) + \quad \stackrel{?}{=} 7$ $\qquad$ $\quad - 3(\quad) \stackrel{?}{=} 11$ $\qquad$ Substitute for x and y.

$= 7$ ✓ $\qquad$ $\neq 11$ ✗

The ordered pair (1, 2) is of the system of of equations.

B. (2, −3)

$5x + y = 7$ $\qquad$ $x - 3y = 11$

$5(\quad) + \quad \stackrel{?}{=} 7$ $\qquad$ $\quad - 3(\quad) \stackrel{?}{=} 11$ $\qquad$ Substitute for x and y.

$= 7$ ✓ $\qquad$ $= 11$ ✓

The ordered pair (2, −3) is of the system of equations.

Example 2

Solve the system of equations.

$$y = x - 4$$
$$y = 2x - 9$$

$y = y$

$y = x - 4$ **$y = 2x - 9$**

$x - 4 = 2x - 9$

Solve the equation to find x.

$x - 4 = \quad 2x - 9$

$-$ ______ $-$ ______ Subtract from both sides.

$- 4 = x - 9$

$+$ ______ $+$ ______ Add to both sides.

$= x$

To find y, substitute for x in one of the original equations.

$y = x - 4 = \quad - 4 =$

The solution is (,) .

Check: Substitute for x and for y in each equation.

$y = x - 4$ $y = 2x - 9$

$\stackrel{?}{=} \quad - 4$ $\stackrel{?}{=} 2(\quad) - 9$

$= \quad$ ✓ $= \quad$ ✓

Try This

1. **Determine if the ordered pair is a solution of the system of equations below.**

$$4x + y = 8$$
$$x - 4y = 12$$

(2, −3)

2. **Solve the system of equations.**

$$y = x - 5$$
$$y = 2x - 8$$

3. **Solve the system of equations.**

$$x + y = 5$$
$$3x + y = -1$$

Chapter 10
Tetra Terms

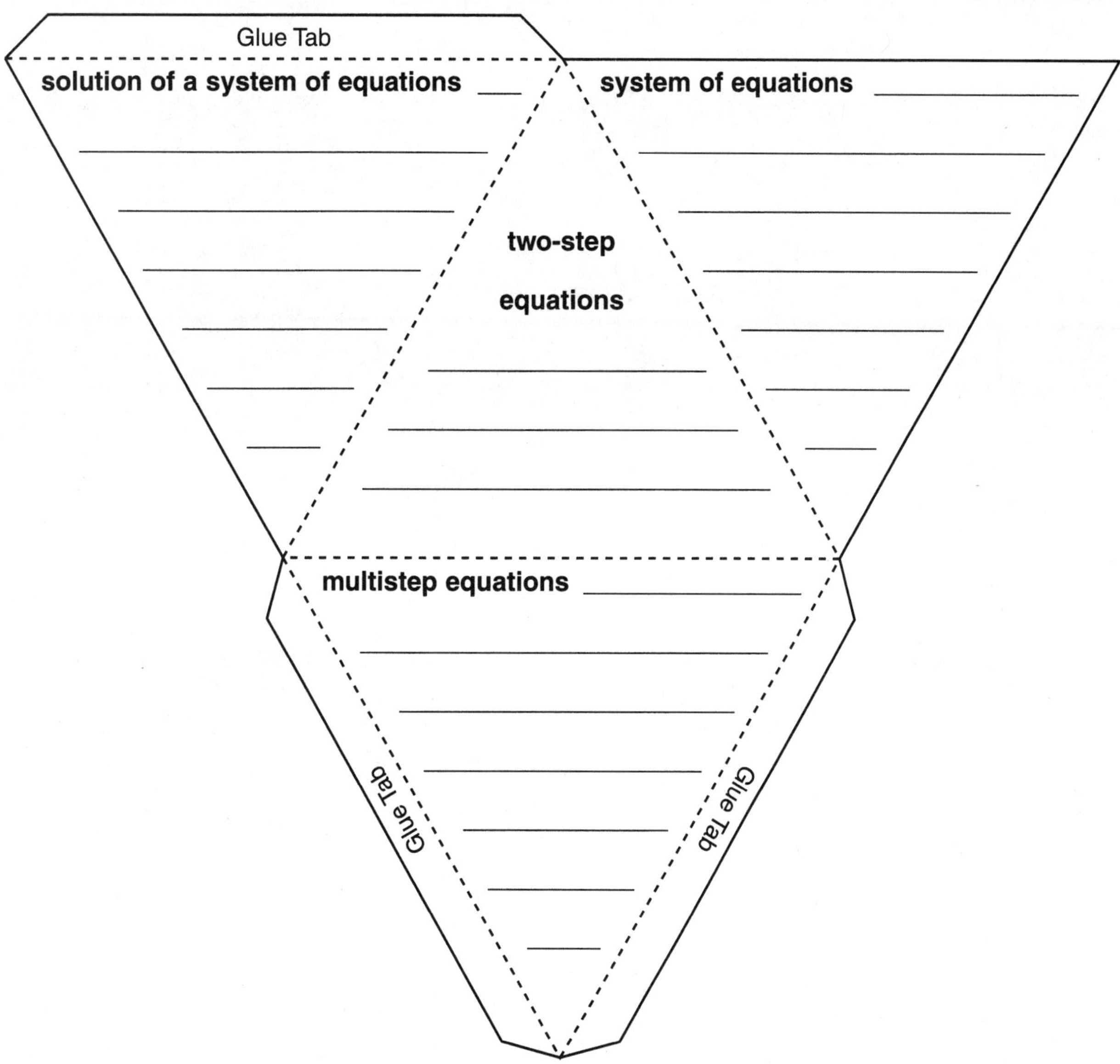

Directions

1. Write the definition of each term on the net of the tetrahedron.
2. Cut out the net.
3. Fold along all dotted lines, and place glue tabs to the inside of the tetrahedron.
4. Join the common edges, and tape or glue the tabs in place.

LESSON **11-1** # Graphing Linear Equations
pp. 540–542

Vocabulary

linear equation (p. 540) ______

______

Additional Examples

Example 1

Graph each equation and tell whether it is linear.

A. $y = 3x - 1$

x	$3x - 1$	y	(x, y)
−2	3 ☐ − 1	☐	☐
−1	3 ☐ − 1	☐	☐
0	3 ☐ − 1	☐	☐
1	3 ☐ − 1	☐	☐
2	3 ☐ − 1	☐	☐

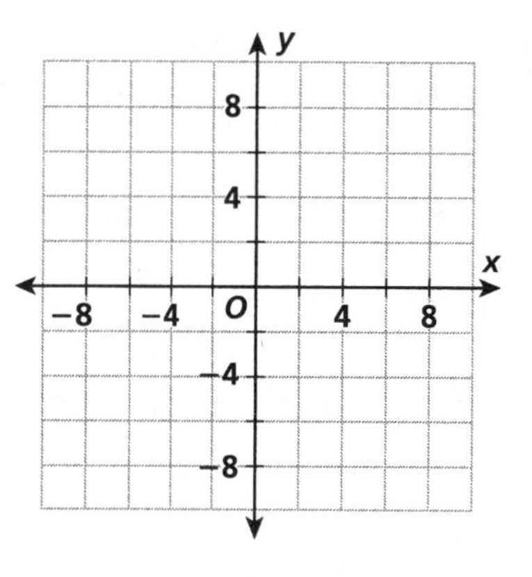

The equation $y = 3x - 1$ is ______ equation because it is the graph of a ______ line and each time x increases by ______ unit, y increases by ______ units.

B. $y = x^3$

x	x^3	y	(x, y)
-2			
-1			
0			
1			
2			

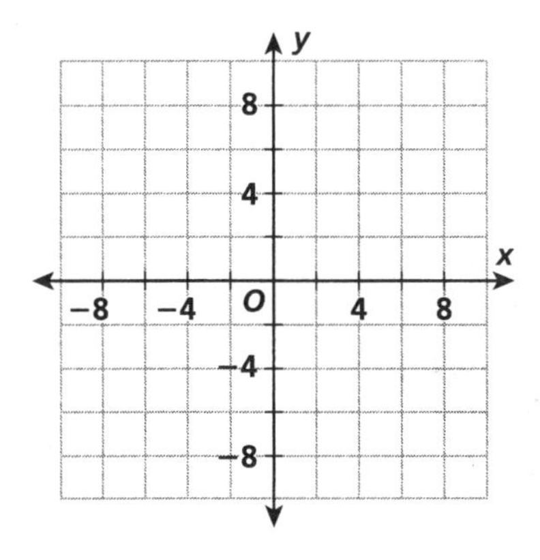

The equation $y = x^3$ is ______ equation because its graph is not a ______ line. Also notice that as x increases by a constant of ______ unit, the change in y is not ______.

Try This

1. Graph the equation and tell whether it is linear.

$y = x$

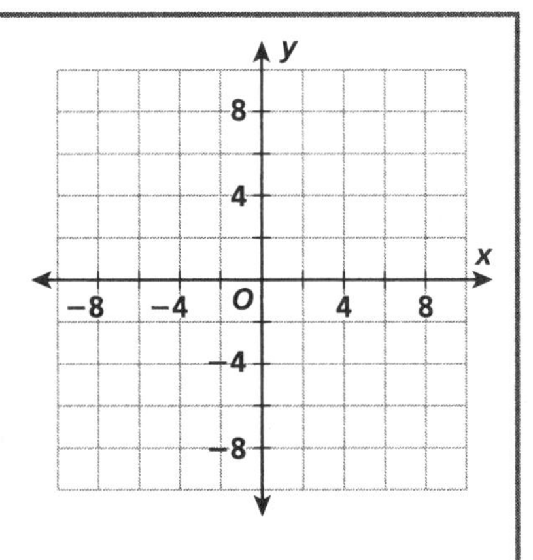

Slope of a Line

pp. 545–547

Additional Examples

Example 2

Use the graph of the line to determine its slope.

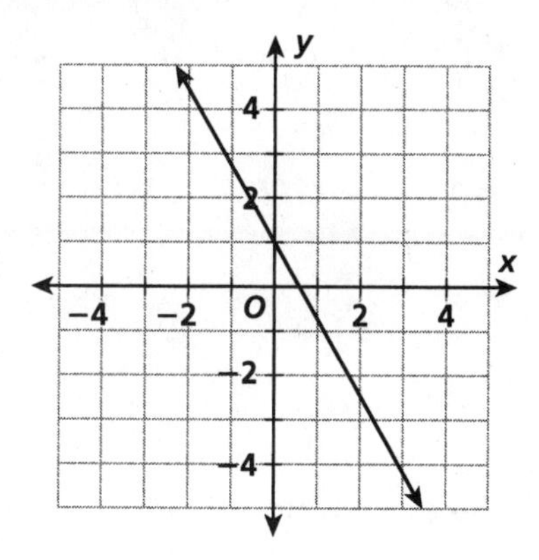

Choose two points on the line: (0, 1) and (3, −4).

Guess by looking at the graph:

$$\frac{\text{rise}}{\text{run}} = \frac{\square}{\square} = -\square$$

Use the slope formula.

Let (3, −4) be (x_1, y_1) and (0, 1) be (x_2, y_2).

$$\frac{y_2 - y_1}{x_2 - x_1} = \frac{\square - (\square)}{\square - \square} = \square = -\square$$

Notice that if you switch (x_1, y_1) and (x_2, y_2), you get the same slope:

Let (0, 1) be (x_1, y_1) and (3,−4) be (x_2, y_2).

$$\frac{y_2 - y_1}{x_2 - x_1} = \frac{\square - \square}{\square - \square} = \square = -\square$$

The slope of the given line is ______.

Example 3

Tell whether the lines passing through the given points are parallel or perpendicular.

A. line 1: (−6, 4) and (2, −5); line 2: (−1, −4) and (8, 4)

slope of line 1: $\frac{y_2 - y_1}{x_2 - x_1} = \frac{\quad - \quad}{\quad - (\quad)} = \quad -$

slope of line 2: $\frac{y_2 - y_1}{x_2 - x_1} = \frac{\quad - (\quad)}{\quad - (\quad)} =$

Line 1 has a slope equal to and line 2 has a slope equal to

, − and are negative of each

other, so the lines are .

B. line 1: (0, 5) and (6, −2); line 2: (−1, 3) and (5, −4)

slope of line 1: $\frac{y_2 - y_1}{x_2 - x_1} = \frac{\quad - \quad}{\quad - (\quad)} = \quad = -$

slope of line 2: $\frac{y_2 - y_1}{x_2 - x_1} = \frac{\quad - \quad}{\quad - (\quad)} = \quad = -$

Both lines have a slope equal to , so the lines are .

Example 3

Graph the line passing through (3, 1) with slope 2.

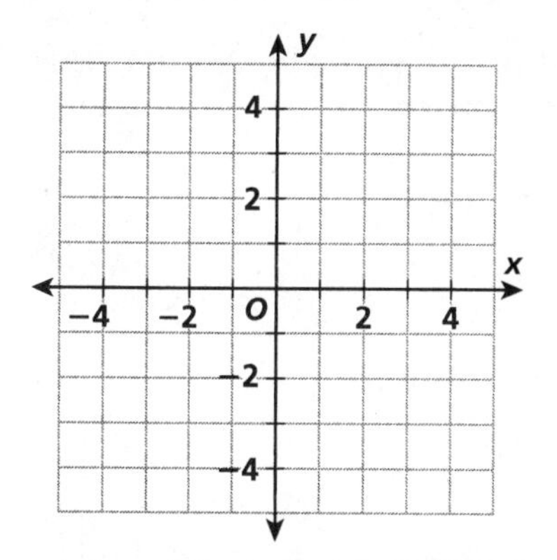

The slope is 2, or $\frac{2}{1}$. So for every 2 units up, you will move right ______ unit, and for every 2 units down, you will move left ______ unit.

Plot the point (3, 1). Then move 2 units up and right ______ unit and plot the point (______). Use a straightedge to connect the two points.

Try This

1. Find the slope of the line that passes through (−4, −6) and (2, 3).

2. Use the graph of the line to determine its slope.

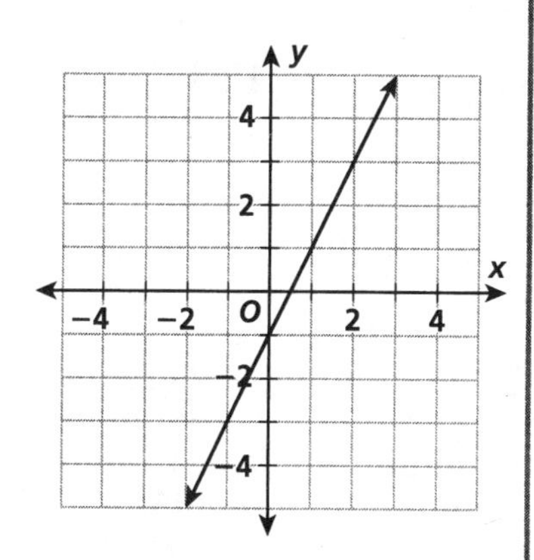

3. Tell whether the lines passing through the given points are parallel or perpendicular.

line 1: (−8, 2) and (0, −7); line 2: (−3, −6) and (6, 2)

LESSON **11-3**

Using Slopes and Intercepts

pp. 550–552

Vocabulary

x-intercept (p. 550)

y-intercept (p. 550)

slope-intercept form (p. 551)

Additional Examples

Example 1

Find the x-intercept and y-intercept of the line $4x - 3y = 12$. Use the intercepts to graph the equation.

Find the x-intercept ($y =$).

$$4x - 3y = 12$$

$$4x - 3(\quad) = 12$$

$$4x = 12$$

$$\frac{4x}{4} = \frac{12}{4}$$

$$x =$$

The x-intercept is   .

Find the y-intercept ($x =$).

$$4x - 3y = 12$$

$$4(\quad) - 3y = 12$$

$$-3y = 12$$

$$\frac{-3y}{-3} = \frac{12}{-3}$$

$$y =$$

The y-intercept is   .

The graph of $4x - 3y = 12$ is the line that crosses the x-axis at the point (,   ) and the y-axis at the point (,   ).

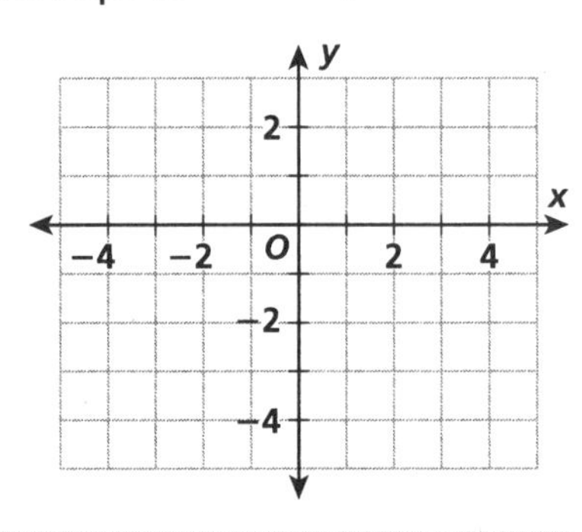

Example 2

Write each equation in slope-intercept form, and then find the slope and *y*-intercept.

A. $2x + y = 3$

$$2x + y = 3$$

$- \_\_\_\_ \quad - \_\_\_\_$ Subtract ____ from both sides.

$$y = 3 - 2x$$

Rewrite to match slope-intercept form.

$y =$ ____ The equation is in ____-intercept form.

$m =$ ____ $b =$ ____

The slope of the line $2x + y = 3$ is ____, and the *y*-intercept is ____.

Example 4

Write the equation of the line that passes through (3, −4) and (−1, 4) in slope-intercept form.

Find the slope.

$\dfrac{y_2 - y_1}{x_2 - x_1} = \dfrac{\_\_\_\_ - (\_\_\_\_)}{\_\_\_\_ - \_\_\_\_} = \_\_\_\_ = \_\_\_\_$ The slope is ____.

Choose either point and substitute it along with the slope into the slope-intercept form.

$y = mx + b$

____ $=$ ____ $+ b$ Substitute −1 for *x*, 4 for *y*, and −2 for *m*.

____ $=$ ____ $+ b$ Simplify.

Solve for *b*.

$4 = \quad 2 + b$

$- \_\_\_\_ \quad - \_\_\_\_$ Subtract from both sides.

$= b$

Write the equation of the line, using for *m* and for *b*.

$y =$

Try This

1. Find the *x*-intercept and *y*-intercept of the line 8*x* – 6*y* = 24. Use the intercepts to graph the equation.

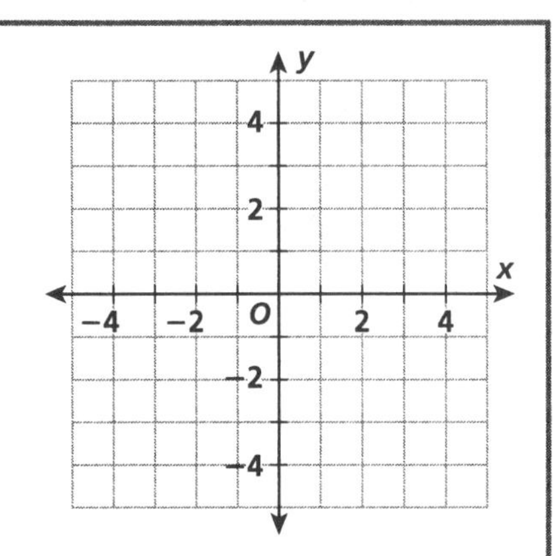

2. Write the equation in slope-intercept form, and then find the slope and *y*-intercept.

$5x + 4y = 8$

3. Write the equation of the line that passes through (1, 2) and (2, 6) in slope-intercept form.

LESSON 11-4

Point-Slope Form

pp. 556–557

Vocabulary

point-slope form (p. 556) ______

Additional Examples

Example 1

Use the point-slope form of the equation to identify a point the line passes through and the slope of the line.

$y - 7 = 3(x - 4)$

$y - y_1 = m(x - x_1)$

$y -$ ____ $=$ ____ $(x -$ ____ $)$ The equation is in ______-slope form.

$m = 3$ Read the value of m from the equation.

$(x_1, y_1) = ($ ____ $)$ Read the point from the equation.

The line defined by $y - 7 = 3(x - 4)$ has slope ____, and passes through the point (____).

Example 2

Write the point-slope form of the equation with the given slope that passes through the indicated point.

A. the line with slope 4 passing through (5, −2)

$y - y_1 = m(x - x_1)$

$[y - (\quad)] = \quad (x - \quad)$ Substitute for x_1, for y_1, and for m.

$y + \quad = \quad (x - \quad)$

The equation of the line with slope 4 that passes through (5, −2) in point-slope form is .

Try This

1. Use the point-slope form of the equation to identify a point the line passes through and the slope of the line.

$y - 2 = \frac{2}{3}(x + 3)$

2. Write the point-slope form of the equation with the given slope that passes through the indicated point.

the line with slope 2 passing through (2, −2)

LESSON **11-5** # Direct Variation
pp. 562–564

Vocabulary

direct variation (p. 562)

constant of proportionality (p. 562)

Additional Examples

Example 1

Determine whether the data set shows direct variation.

Make a graph that shows the relationship between Adam's age and his length.

Adam's Growth Chart				
Age (mo)	3	6	9	12
Length (in.)	22	24	25	27

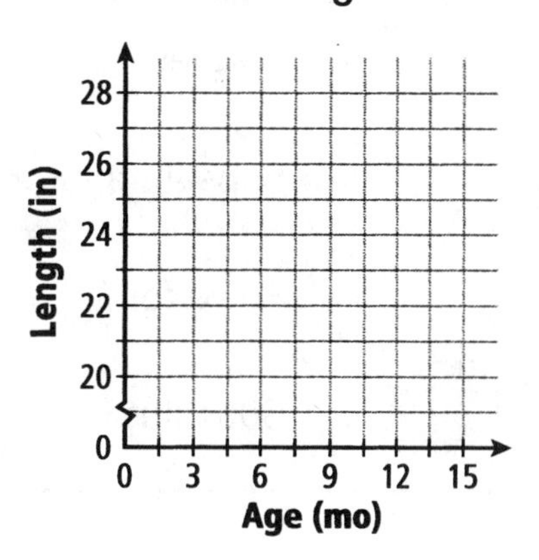

You can also compare ratios to see if a direct ______ occurs.

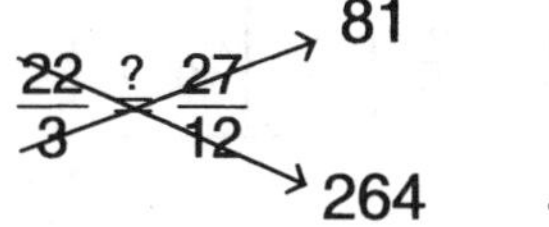

______ $\neq$ ______

The ratios are ______.

The relationship of the data is ______.

Example 2

Find each equation of direct variation, given that y varies directly with x.

A. y is 54 when x is 6

$y = kx$ — y varies with x.

$= k \cdot$ — Substitute for x and y.

$= k$ — Solve for k.

$y =$ — Substitute for k in the original equation.

B. x is 12 when y is 15

$y = kx$ — y varies with x.

$= k \cdot$ — Substitute for x and y.

$= k$ — Solve for k.

$y =$ — Substitute for k in the original equation.

Try This

1. Determine whether the data set shows direct variation.

Kyle's Basketball Shots			
Distance (ft)	20	30	40
Number of Baskets	5	3	0

2. Find the equation of direct variation, given that y varies directly with x.

y is 7 when x is 3

LESSON 11-6

Graphing Inequalities in Two Variables *pp. 567–569*

Vocabulary

boundary line (p. 567) ______

linear inequality (p. 567) ______

Additional Examples

Example 1

Graph each inequality.

A. $y < x - 1$

First graph the boundary line $y = x - 1$. Since no points that are on the line are solutions of $y < x - 1$, make the line ______. Then determine on which side of the line the solutions lie.

(0, 0) Test a point not on the line.

$y < x - 1$

$\square \overset{?}{<} \square - 1$ Substitute $\square$ for x and $\square$ for y.

$\square \overset{?}{<} \square$

Since $0 < -1$ is not true, (0, 0) is not a solution of $y < x - 1$. Shade the side of the line that does not include (0, 0).

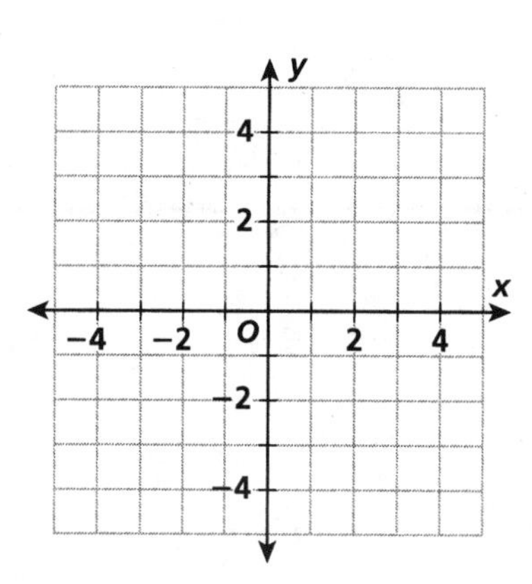

B. $y \geq 2x + 1$

First graph the boundary line $y = 2x + 1$. Since points that are on the line

are solutions of $y \geq 2x + 1$, make the line . Then shade the part

of the coordinate plane in which the rest of the solutions of $y \geq 2x + 1$ lie.

(0, 4) Choose any point not on the line.

$y \geq 2x + 1$

$4 \overset{?}{\geq} 0 + 1$ Substitute for x and for y.

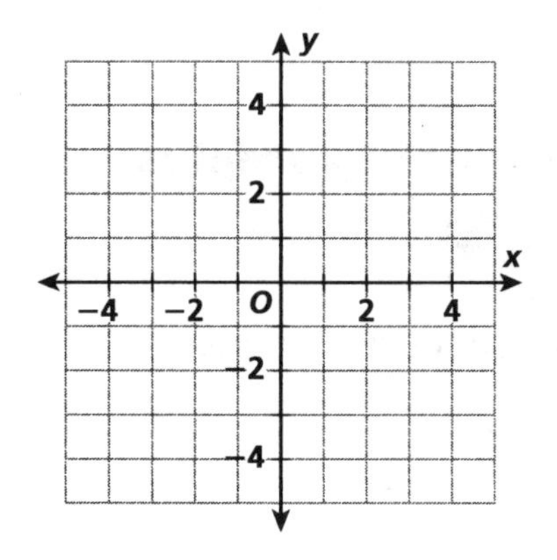

Since $4 \geq 1$ is true, (0, 4) is a solution of $y \geq 2x + 1$. Shade the side of the line that includes (0, 4).

Try This

1. Graph the inequality.

$y < x - 4$

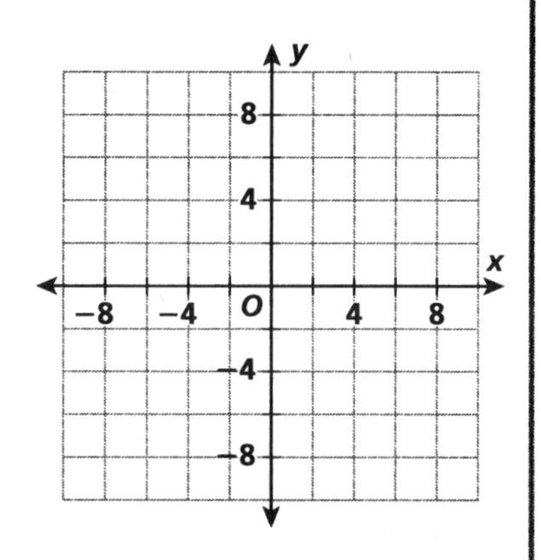

LESSON **11-7**

Lines of Best Fit

pp. 572–573

Additional Examples

Example 1

Plot the data and find a line of best fit.

x	4	7	3	8	8	6
y	4	5	2	6	7	4

Plot the data points and find the mean of the x- and y-coordinates.

$$x_m = \frac{4 + 7 + 3 + 8 + 8 + 6}{\square} = \square$$

$$y_m = \frac{4 + 5 + 2 + 6 + 7 + 4}{\square} = \square$$

$$(x_m, y_m) = (\quad\quad)$$

Draw a line through $(6, 4\frac{2}{3})$ that best represents the data.

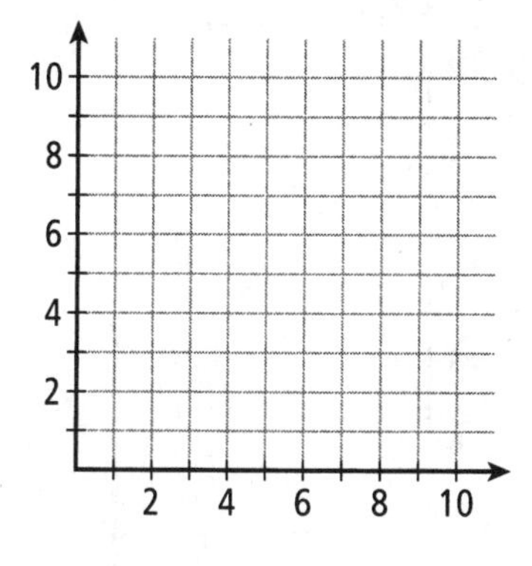

Estimate and plot the coordinates of another point on that line, such as (8, 6). Find the equation of the line.

$m = \frac{\quad - \quad}{\quad - \quad} = \quad = \quad$ Find the slope.

$y - y_1 = m(x - x_1)$ Use - form.

$y - \quad = \quad (x - \quad)$ Substitute.

$y - 4\frac{2}{3} = \frac{2}{3}x - 4$

$y = \quad x +$

The equation of a line of best fit is .

Try This

1. Plot the data and find a line of best fit.

x	−1	0	2	6	−3	8
y	−1	0	3	7	−7	4

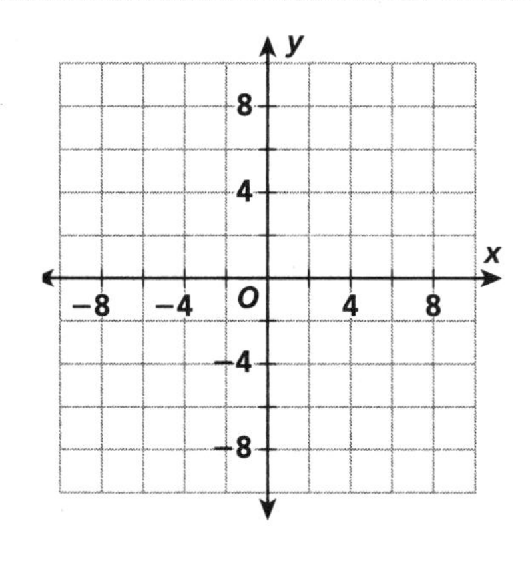

Chapter 11
Property Prism

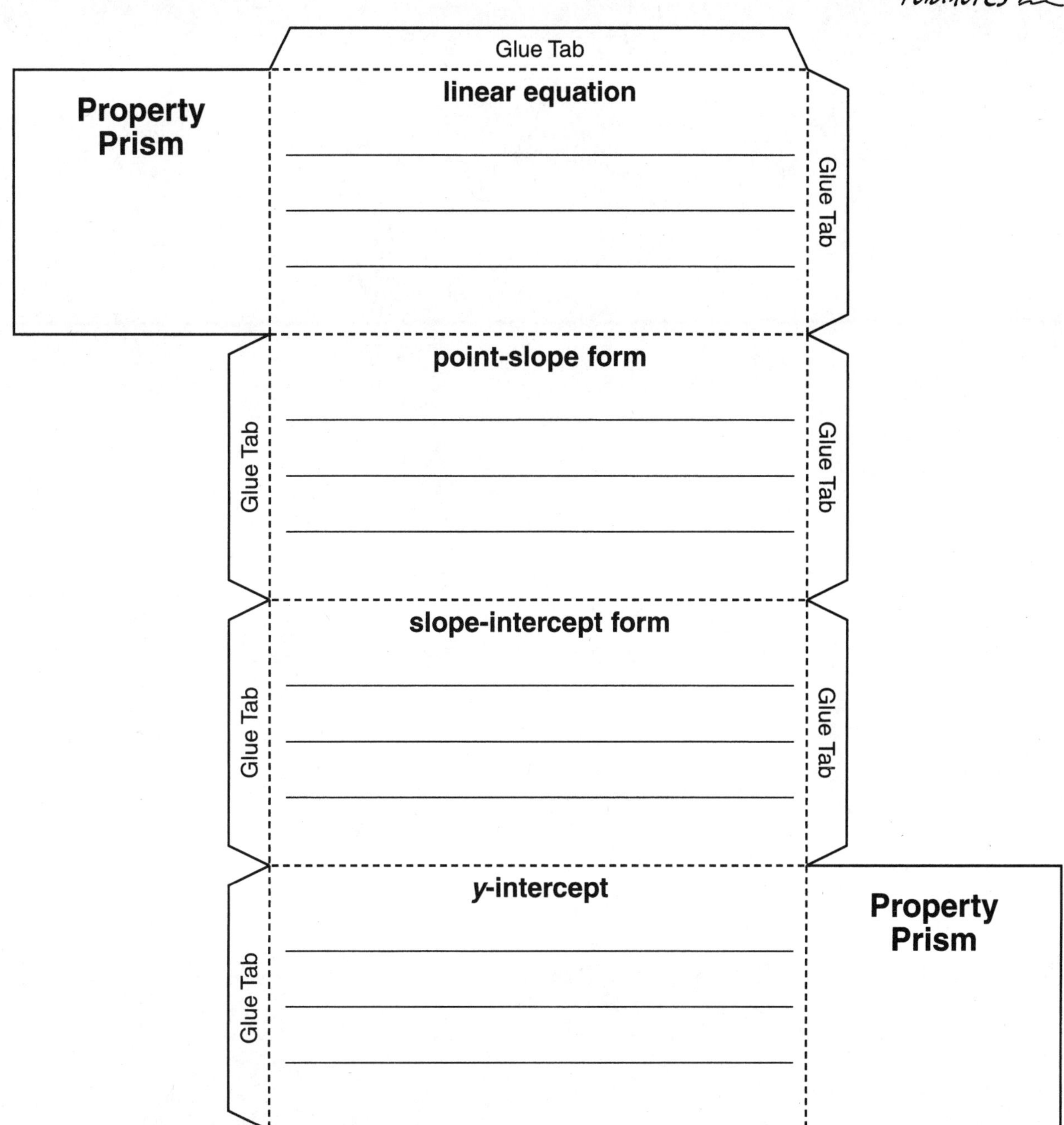

Directions

1. Write the definition of each term on the net.
2. Cut out the net of the rectangular prism.
3. Fold along all dotted lines, and place glue tabs to the inside of the prism.
4. Join the common edges, and tape or glue the tabs in place.

LESSON **12-1**

Arithmetic Sequences

pp. 590–592

Know-It Notes

Vocabulary

sequence (p. 590) ______

term (p. 590) ______

arithmetic sequence (p. 590) ______

common difference (p. 590) ______

Additional Examples

Example 1

Determine if each sequence could be arithmetic. If so, give the common difference.

A. 5, 8, 11, 14, 17, . . .

5, 8, 11, 14, 17, . . .

Find the difference of each term and the term before it.

The sequence could be ______ with a common difference of ______.

B. 1, 3, 6, 10, 15, . . .

1, 3, 6, 10, 15, . . .

Find the difference of each term and the term before it.

The sequence is not ______.

Example 2

Find the given term in each arithmetic sequence.

A. 10th term: 1, 3, 5, 7, . . .

$$a_n = a_1 + (n - 1)d$$

a = + (− 1)

a =

B. 18th term: 100, 93, 86, 79, . . .

$$a_n = a_1 + (n - 1)d$$

a = + (− 1)( )

a =

C. 21st term: 25, 25.5, 26, 26.5, . . .

$$a_n = a_1 + (n - 1)d$$

a = + (− 1)( )

a =

Try This

1. Determine if the sequence could be arithmetic. If so, give the common difference.

11, 22, 33, 44, 55, . . .

2. Find the given term in the arithmetic sequence.

15th term: 1, 3, 5, 7, . . .

LESSON **12-2**

Geometric Sequences

pp. 595–597

Vocabulary

geometric sequence (p. 595) ______________________

common ratio (p. 595) ______________________

Additional Examples

Example 1

Determine if each sequence could be geometric. If so, give the common ratio.

A. 1, 5, 25, 125, 625, . . .

1, 5, 25, 125, 625, . . . Divide each term by the term before it.

The sequence could be a ______ with a common ratio of ___.

B. 1, 3, 9, 12, 15, . . .

1, 3, 9, 12, 15, . . . Divide each term by the term before it.

The sequence is not ______.

C. 81, 27, 9, 3, 1, . . .

81, 27, 9, 3, 1, . . . Divide each term by the term before it.

The sequence could be ______ with a common ratio of $\frac{1}{3}$.

Example 2

Find the given term in each geometric sequence.

A. 11th term: $-2, 4, -8, 16, \ldots$

$r = \underline{\qquad} =$

$a_n = a_1 r^{n-1}$

$a \qquad = \qquad (\qquad) \qquad = -2(1024) =$

B. 9th term: 100, 70, 49, 34.3, . . .

$r = \underline{\qquad} =$

$a_n = a_1 r^{n-1}$

$a \qquad = \qquad (\qquad) \qquad = 100(0.05764801) =$

Try This

1. Determine if the sequence could be geometric. If so, give the common ratio.

2, 4, 12, 24, 96, . . .

2. Find the given term in the geometric sequence.

11th term: 100, 70, 49, 34.3, . . .

LESSON **12-3**

Other Sequences

pp. 601–603

Vocabulary

first differences (p. 601) ______________________________

____

second differences (p. 601) ______________________________

____

Fibonacci sequence (p. 603) ______________________________

____

____

Additional Examples

Example 1

Use first and second differences to find the next three terms in each sequence.

A. 1, 8, 19, 34, 53, . . .

Sequence	1	8	19	34	53			
1st Differences								
2nd Differences								

The next three terms are ____, ____, ____.

B. 12, 15, 21, 32, 50, . . .

Sequence	12	15	21	32	50			
1st Differences								
2nd Differences								

The next three terms are ____, ____, ____.

Example 2

Give the next three terms in each sequence, using the simplest rule you can find.

A. 1, 2, 1, 1, 2, 1, 1, 1, 2, . . .

One possible rule is to have 1 in front of the 1st 2, 1s in front of the 2nd 2, 1s in front of the 3rd 2, and so on.

The next three terms are , , .

B. $\frac{2}{5}, \frac{3}{7}, \frac{4}{9}, \frac{5}{11}, \frac{6}{13}, \ldots$

One possible rule is to 1 to the numerator and 2 to the denominator of the previous term. This could be written as the algebraic rule.

$a_n = \frac{n+1}{2n+1}$

The next three terms are , , .

C. 1, 11, 6, 16, 11, 21, . . .

A rule for the sequence could be to start with 1 and use the pattern of adding , subtracting to get the next two terms.

The next three terms are , , .

D. 1, −2, 3, −4, 5, −6, . . .

A rule for the sequence could be the set of counting numbers with every even number being multiplied by .

The next three terms are , , .

Example 3

Find the first five terms of the sequence defined by $a_n = n(n - 2)$.

$a_1 = \square(\square - 2) = \square$

$a_2 = \square(\square - 2) = \square$

$a_3 = \square(\square - 2) = \square$

$a_4 = \square(\square - 2) = \square$

$a_5 = \square(\square - 2) = \square$

The first five terms are $\square$, $\square$, $\square$, $\square$, $\square$.

Try This

1. Use first and second differences to find the next three terms in each sequence.

2, 4, 10, 20, 34, . . .

2. Give the next three terms in the sequence, using the simplest rule you can find.

1, 2, 3, 5, 7, 11, . . .

LESSON **12-4**

Functions

pp. 608–610

Vocabulary

function (p. 608)

input (p. 608)

output (p. 608)

domain (p. 608)

range (p. 608)

function notation (p. 609)

Additional Examples

Example 1

Make a table and a graph of $y = 3 - x^2$.

Make a table of inputs and outputs. Use the table to make a graph.

x	$3 - x^2$	y
−2		
−1		
0		
1		
2		

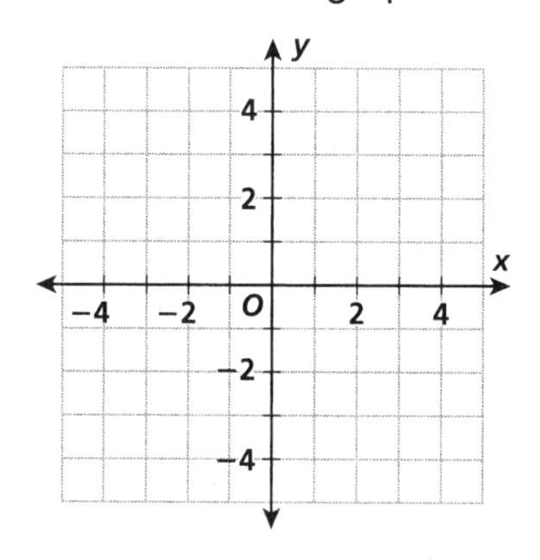

Example 2

Determine if each relationship represents a function.

B.

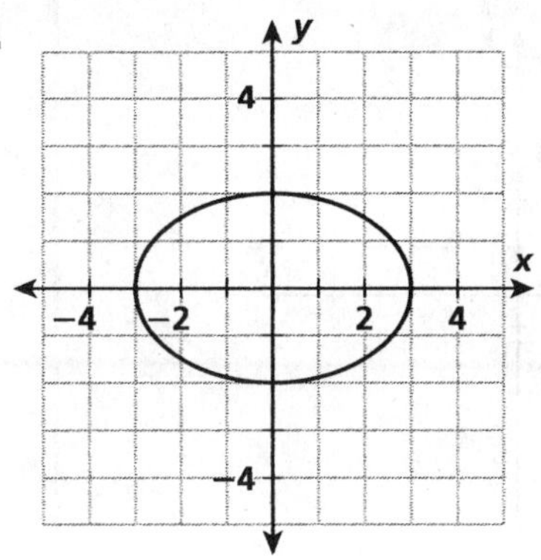

The input $x = 0$ has ______ outputs,

$y =$ ______ and $y =$ ______. Other x-values also have more than one y-value. The relationship is ______.

C. $y = x^3$

Make an input-output table and use it to graph $y = x^3$.

x	y
−2	
−1	
0	
1	
2	

y
2
x
−2
O
2
−2

Each input x has only one output y. The relationship

______.

Try This

1. Make a table and a graph of $y = x + 1$.

x	$x + 1$	y
-1		
0		
1		
2		

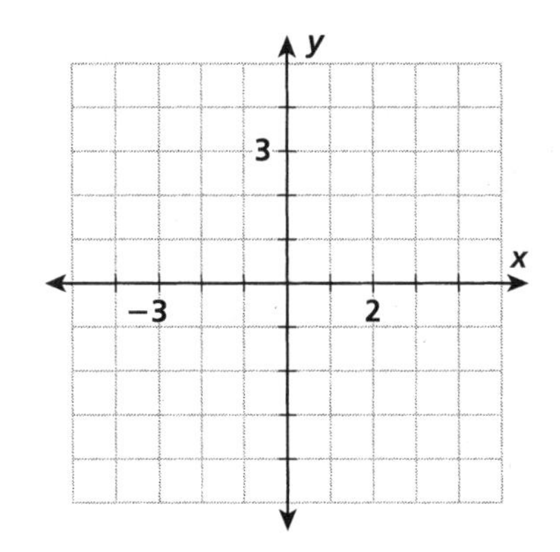

2. Determine if the relationship represents a function.

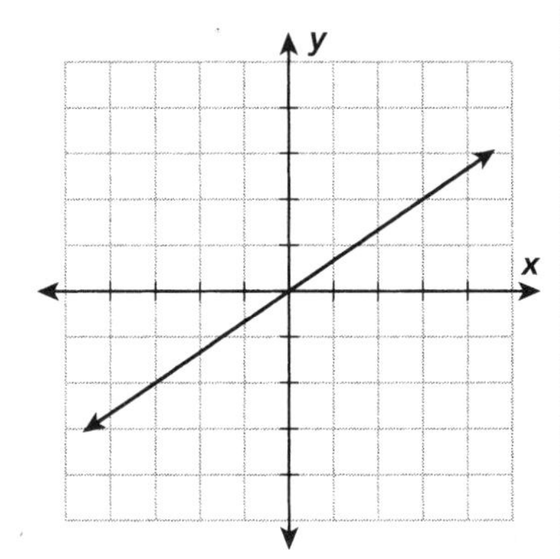

3. For the function, find $f(0)$, $f(-1)$, and $f(1)$.

$y = 2x + 2$

LESSON **12-5**

Linear Functions

pp. 613–614

Vocabulary

linear function (p. 613) ______________________________

____

Additional Examples

Example 1

Write the rule for the linear function.

Use the equation $f(x) = mx + b$. To find b, identify the y-intercept from the graph.

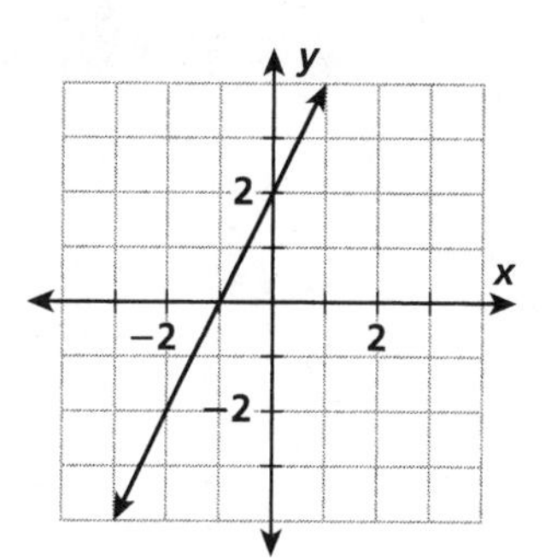

$b =$ ☐

$f(x) = mx +$ ☐

Locate another point on the graph, such as (1, 4). Substitute the x- and y-values of the point into the equation, and solve for m.

$$f(x) = mx + 2$$
$$4 = m(1) + 2 \qquad (x, y) = (1, 4)$$
$$4 = m + 2$$

$-$ ☐ $\qquad -$ ☐

☐ $= m$

The rule is ______________.

Example 2

Write the rule for each linear function.

A. The *y*-intercept can be identified from the table as $b = f(0) = 1$. Substitute the *x*- and *y*-values of the point (1, –1) into the equation $f(x) = mx + 1$, and solve for *m*.

x	*y*
−2	5
−1	3
0	1
1	−1

$f(x) = mx + 1$

$\quad = m(\quad) + 1$

$\quad = m + 1$

$-\_\_\_\_ \qquad -\_\_\_\_$

$\quad = m$

The rule is ______________.

Try This

1. Write the rule for the linear function.

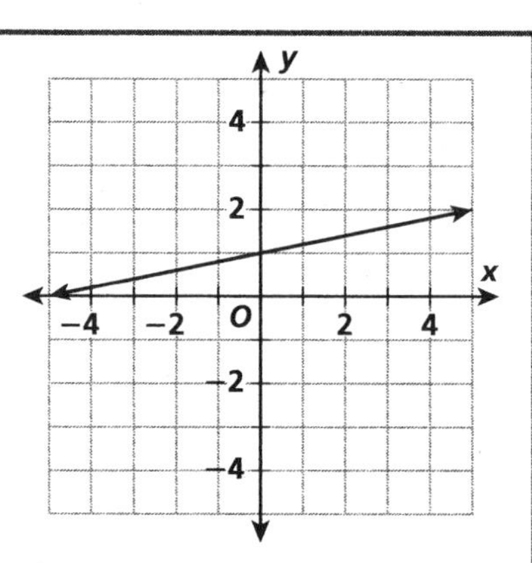

2. Write the rule for the linear function.

x	*y*
0	0
−1	1
1	−1
2	−2

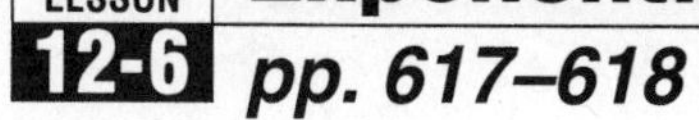

Exponential Functions

LESSON 12-6 pp. 617–618

Vocabulary

exponential function (p. 617) ______

exponential growth (p. 618) ______

exponential decay (p. 618) ______

Additional Examples

Example 1

Create a table for each exponential function, and use it to graph the function.

A. $f(x) = 3 \cdot 2^x$

x	y
−2	
−1	
0	
1	
2	

$3 \cdot 2^{\square} = 3 \cdot \square$

$3 \cdot 2^{\square} = 3 \cdot \square$

$3 \cdot 2^{\square} = 3 \cdot \square$

$3 \cdot 2^{\square} = 3 \cdot \square$

$3 \cdot 2^{\square} = 3 \cdot \square$

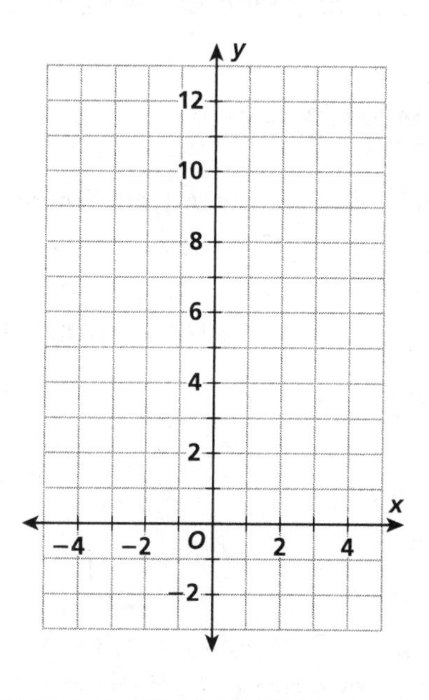

Example 2

A bacterial culture contains 5000 bacteria, and the number of bacteria doubles each day. How many bacteria will be in the culture after a week?

Day	Mon	Tue	Wed	Thu
Number of days x	0	1	2	3
Number of bacteria $f(x)$	5000	10,000	20,000	40,000

$f(x) = p \cdot a^x$

$f(x) = \quad \cdot a^x$ $\qquad$ $f(0) = p$

$f(x) = \quad \cdot \quad ^x$ $\qquad$ $f(1) = 5000 \cdot a^1 = 10{,}000$, so $a = 2$.

A week is 7 days so let $x = 7$.

$f(7) = 5000 \cdot 2 \quad = $ $\qquad$ Substitute for x.

If the number of bacteria doubles each day, there will be

bacteria in the culture after a week.

Example 3

Bohrium-267 has a half-life of 15 seconds. Find the amount that remains from a 16 mg sample of this substance after 2 minutes.

Seconds	0	15	30	45
Number of Half-lives x	0	1	2	3
Bohrium-267 $f(x)$ (mg)	16	8	4	2

$f(x) = p \cdot a^x$

$f(x) = \quad \cdot a^x$ $\qquad$ $f(0) = p$

$f(x) = \quad \cdot \left(\quad \right)^x$ $\qquad$ $f(1) = 16 \cdot a^1 = 8$, so $a = \left(\frac{1}{2}\right)$.

Since 2 minutes = 120 seconds, divide 120 seconds by 15 seconds to find the number of half-lives: $x = 8$.

$f(8) = 16 \cdot \left(\frac{1}{2}\right)$ $\qquad$ Substitute for x.

There is mg of Bohrium-267 left after 2 minutes.

Try This

1. Create a table for the exponential function, and use it to graph the function.

$f(x) = 2^x$

x	y
−2	
−1	
0	
1	
2	

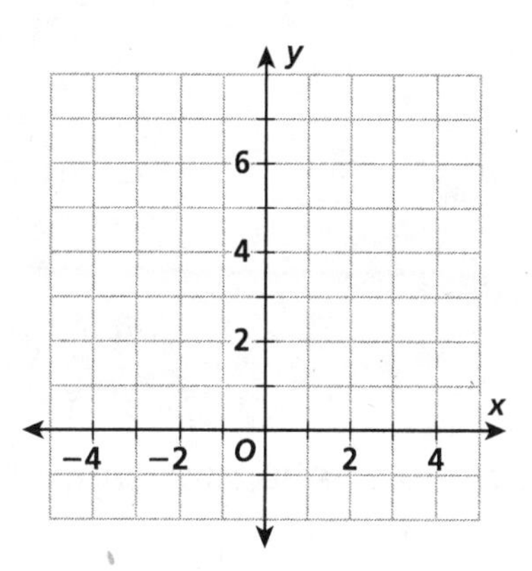

2. Robin invested $300 in an account that will double her balance every 4 years. Write an exponential function to calculate her account balance. What will her account balance be in 20 years?

Year	2003	2007	2011	2015
Every 4 years x	0	1	2	3
Account balance $f(x)$	300	600	1200	2400

3. If an element has a half-life of 25 seconds. Find the amount that remains from a 8 mg sample of this substance after 3 minutes.

Seconds	0	25	50	75
Number of Half-lives x	0	1	2	3
Element (mg)	8	4	2	1

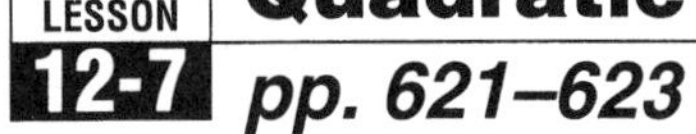

Quadratic Functions
pp. 621–623

Vocabulary

quadratic function (p. 621)

parabola (p. 621)

Additional Examples

Example 1

Create a table for each quadratic function, and use it to make a graph.

A. $f(x) = x^2 + 1$

x	$f(x) = x^2 + 1$
−2	
−1	
0	
1	
2	

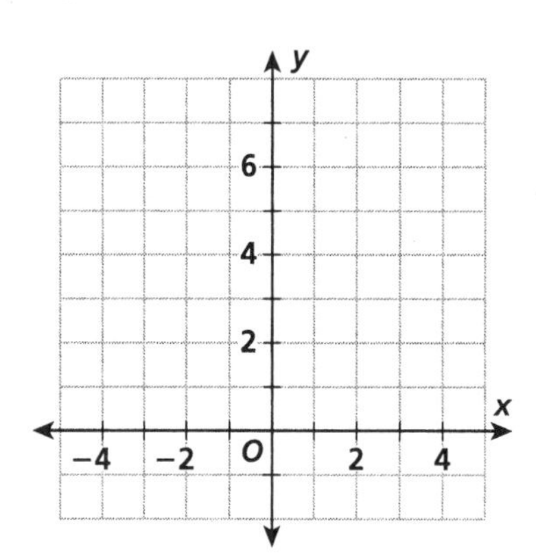

Plot the points and connect them with a smooth curve.

Example 2

Create a table for each quadratic function, and use it to make a graph.

A. $f(x) = (x - 2)(x + 3)$

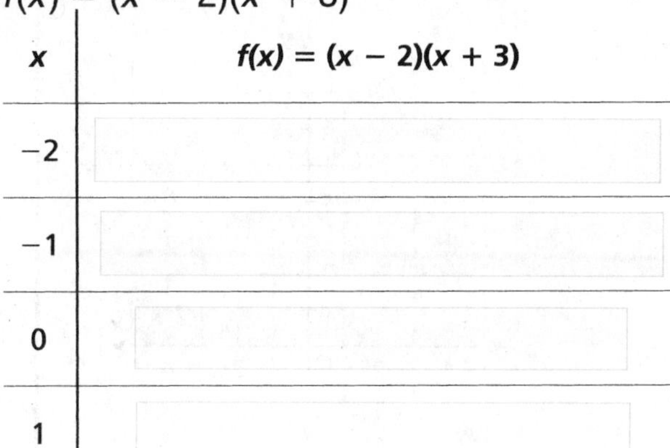

x	$f(x) = (x - 2)(x + 3)$
−2	
−1	
0	
1	
2	

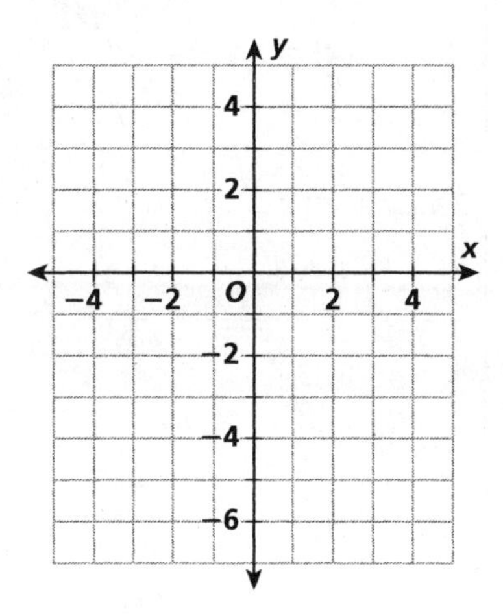

The parabola crosses the x-axis at $x =$ ____ and $x =$ ____.

Plot the points and connect them with a smooth curve.

B. $f(x) = (x - 1)(x + 4)$

x	$f(x) = (x - 1)(x + 4)$
−2	
−1	
0	
1	
2	

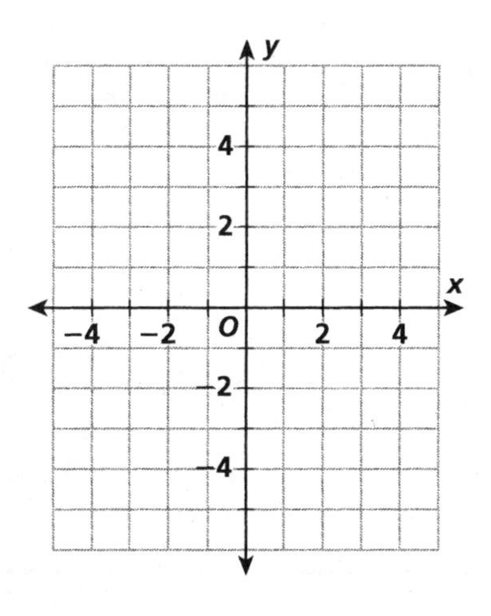

The parabola crosses the x-axis at $x =$ ____ and $x =$ ____.

Plot the points and connect them with a smooth curve.

Try This

1. Create a table for the quadratic function, and use it to make a graph.

$f(x) = x^2 - 1$

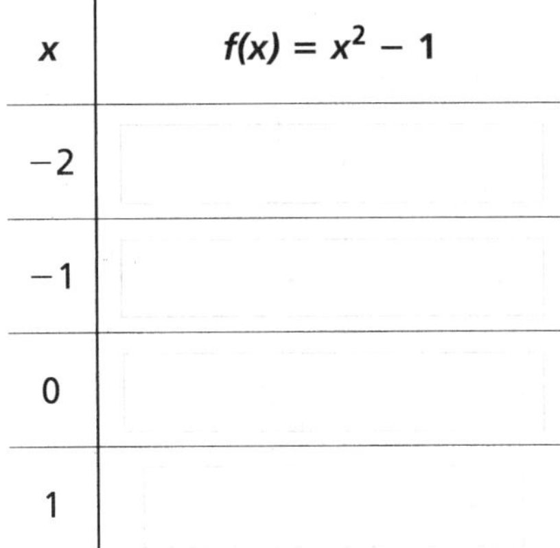

x	$f(x) = x^2 - 1$
−2	
−1	
0	
1	
2	

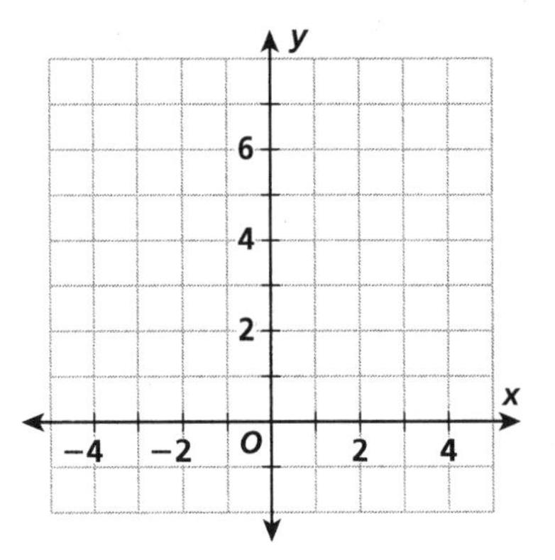

2. Create a table for the quadratic function, and use it to make a graph.

$f(x) = (x - 1)(x + 1)$

x	$f(x) = (x - 1)(x + 1)$
−2	
−1	
0	
1	
2	

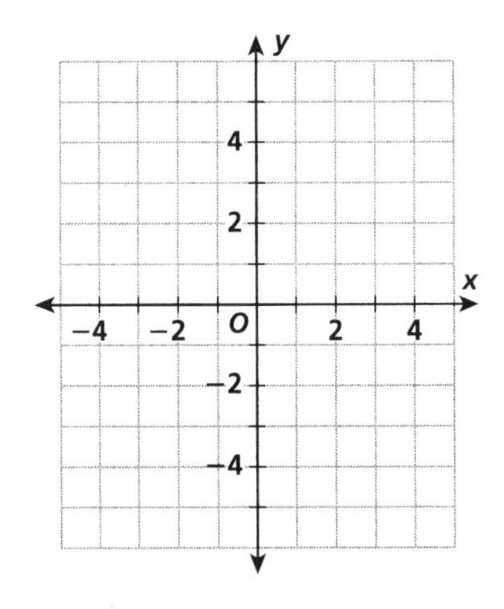

LESSON 12-8

Inverse Variation

pp. 628–629

Vocabulary

inverse variation (p. 628) ______________

____

____

Additional Examples

Example 1

Tell whether each relationship is an inverse variation.

A. The table shows how 24 cookies can be divided equally among different numbers of students.

Number of Students	2	3	4	6	8
Number of Cookies	12	8	6	4	3

$2(12) =$ ____; $3(8) =$ ____; $4(6) =$ ____; $6(4) =$ ____; $8(3) =$ ____

$xy =$ ____ The product is always the same.

The relationship is an ____ variation: $y = \frac{\square}{x}$.

B. The table shows the number of cookies that have been baked at different times.

Number of Cookies	12	24	36	48	60
Time (min)	15	30	45	60	75

$12(15) =$ ____; $24(30) =$ ____ The product is not always the same.

The relationship is not an ____ variation.

Example 2

Graph each inverse variation function.

A. $f(x) = \frac{4}{x}$

x	y
−4	
−2	
−1	
$-\frac{1}{2}$	
$\frac{1}{2}$	
1	
2	
4	

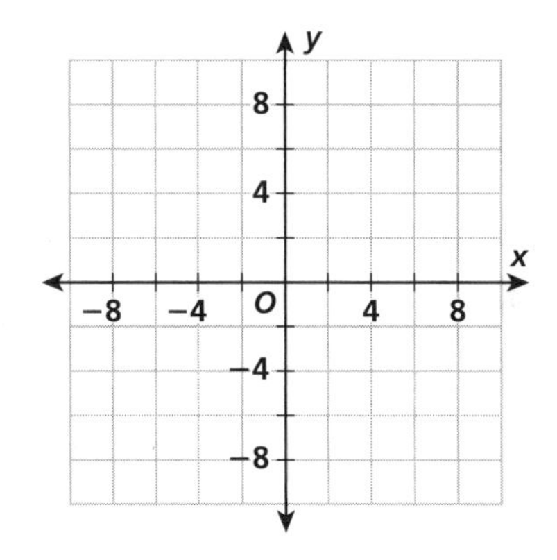

Try This

1. Tell whether the relationship is an inverse variation.

x	30	20	15	12	10
y	2	3	4	5	6

2. Graph the inverse variation function.

$f(x) = -\frac{4}{x}$

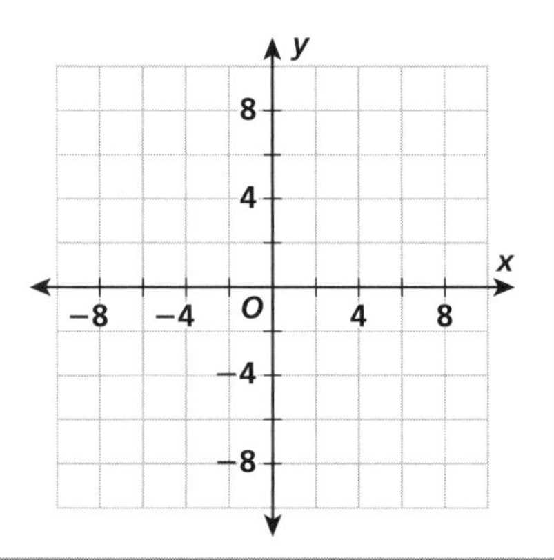

Chapter 12
Möbius Mobile

arithmetic sequence ____________
exponential function ____________
Fibonacci sequence ____________
function ____________
function notation ____________
geometric sequence ____________
linear function ____________
quadratic function ____________

Directions

1. Cut each strip from the page before writing the definition.
2. Begin the definition on the same line as the word.
3. If a second line is needed, flip the strip toward you and continue on the top line. If a third line is needed, flip the strip back to the original side and continue on the next line. Continue this process until finished.
4. Hold the strip with the original side in view. Bring the two ends toward each other so the labels on the tabs are visible.
5. Flip the tab on the right and place it over tab A such that neither tab is visible.
6. Tape them in place.
7. Use string and the strips to build a Möbius mobile.

Chapter 12
Möbius Mobile

Flip		Tab A
Flip		Tab A
Flip		Tab A
Flip		Tab A
Flip		Tab A
Flip		Tab A
Flip		Tab A
Flip		Tab A

LESSON **13-1**

Polynomials

pp. 644–645

Vocabulary

monomial (p. 644)

polynomial (p. 644)

binomial (p. 644)

trinomial (p. 644)

degree of a polynomial (p. 645)

Additional Examples

Example 1

Determine whether each expression is a monomial.

A. $\sqrt{2} \cdot x^3y^4$

3 and 4 are ______ numbers

B. $3x^3\sqrt{y}$

y does not have an ______ that is a whole number

Example 3

Find the degree of each polynomial.

A. $5x - 2x^2 + 6$

$5x \quad - \quad 2x^2 \quad + \quad 6$

Degree　　Degree　　Degree

The degree of $5x - 2x^2 + 6$ is ___.

B. $-3x^4 + 8x^5 - 4x^6$

$-3x^4 \quad + \quad 8x^5 \quad - \quad 4x^6$

Degree　　Degree　　Degree

The degree of $-3x^4 + 8x^5 - 4x^6$ is ___.

Try This

1. Determine whether each expression is a monomial.

A. $2w \cdot p^3y^8$

B. $9t^{3.2}z$

2. Classify each expression as a monomial, a binomial, a trinomial, or not a polynomial.

A. $4x^2 + 7z^4$

B. $1.3x^{2.5} + 4y$

3. Find the degree of the polynomial.

A. $x + 4x^4 + 2y$

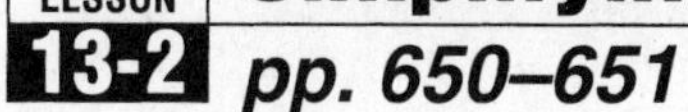

Simplifying Polynomials
pp. 650–651

Additional Examples

Example 1

Identify the like terms in each polynomial.

A. $5x^3 + y^2 + 2 - 6y^2 + 4x^3$

$5x^3 + y^2 + 2 - 6y^2 + 4x^3$ Identify like terms.

Like terms: $5x^3$ and ______, y^2 and ______

B. $3a^3b^2 + 3a^2b^3 + 2a^3b^2 - a^3b^2$

$3a^3b^2 + 3a^2b^3 + 2a^3b^2 - a^3b^2$ Identify ______ terms.

Like terms: ______, ______, and ______

C. $7p^3q^2 + 7p^2q^3 + 7pq^2$

$7p^3q^2 + 7p^2q^3 + 7pq^2$ Identify ______.

There are ______ like terms.

Example 2

Simplify.

A. $4x^2 + 2x^2 + 7 - 6x + 9$

$4x^2 + 2x^2 - 6x + 7 + 9$ Arrange in descending order.

$4x^2 + 2x^2 - 6x + 7 + 9$ Identify like terms.

______

Combine coefficients:

$4 + 2 = 6$ and $7 + 9 = 16$

B. $3n^5m^4 - 6n^3m + n^5m^4 - 8n^3m$

$3n^5m^4 + n^5m^4 - 6n^3m - 8n^3m$ Arrange in descending order.

$3n^5m^4 + n^5m^4 - 6n^3m - 8n^3m$ Identify like terms.

Combine coefficients:

$3 + 1 = 4$ and $-6 - 8 = -14$

Example 3

Simplify.

A. $3(x^3 + 5x^2)$

$3(x^3 + 5x^2)$ Distributive Property

$3 \cdot x^3 + 3 \cdot 5x^2$

B. $-4(3m^3n + 7m^2n) + m^2n$

$-4(3m^3n + 7m^2n) + m^2n$ Distributive Property

$-4 \cdot 3m^3n - 4 \cdot 7m^2n + m^2n$

$-12m^3n - 28m^2n + m^2n$

Combine like terms.

Try This

1. Identify the like terms in each polynomial.

A. $4y^4 + y^2 + 2 - 8y^2 + 2y^4$

B. $7n^4r^2 + 3a^2b^3 + 5n^4r^2 + n^4r^2$

2. Simplify.

A. $2x^3 + 5x^3 + 6 - 4x + 9$

B. $2n^5p^4 - 7n^6p + n^5p^4 - 9n^6p$

3. Simplify by using the Distributive Property.

A. $2(x^3 + 5x^2)$

B. $-2(6m^3p + 8m^2p) + m^2p$

LESSON 13-3

Adding Polynomials

pp. 656–657

Additional Examples

Example 1

Add.

A. $(5x^3 + x^2 + 2) + (4x^3 + 6x^2)$

$5x^3 + x^2 + 2 + 4x^3 + 6x^2$ Associative Property

$x^3 + \quad x^2 + 2$ Combine terms.

B. $(6x^3 + 8y^2 + 5xy) + (4xy - 2y^2)$

$6x^3 + 8y^2 + 5xy + 4xy - 2y^2$ Associative Property

Combine terms.

Example 2

Add.

A. $(4x^2 + 2x + 11) + (2x^2 + 6x + 9)$

$4x^2 + 2x + 11$

$+\ 2x^2 + 6x + \ 9$ Place like terms in columns.

Combine terms.

B. $(3mn^2 - 6m + 6n) + (5mn^2 + 2m - n)$

$3mn^2 - 6m + 6n$

$+\ 5mn^2 + 2m - \ n$ Place like in columns.

Combine terms.

C. $(-x^2y^2 + 5x^2) + (-2y^2 + 2) + (x^2 + 8)$

$$\begin{array}{lllll} & -x^2y^2 & + 5x^2 & & \\ & & & -2y^2 & + 2 \\ + & & x^2 & & + 8 \\ \hline \end{array}$$

Place like ______ in columns.

Combine ______ terms.

Try This

Add.

1. $(3y^4 + y^2 + 6) + (5y^4 + 2y^2)$

2. $(4mn^2 + 6m + 2n) + (2mn^2 - 2m - 2n)$

3. $(x^2y^2 - 5x^2) + (2y^2 - 2) + (x^2)$

LESSON **13-4**

Subtracting Polynomials

pp. 660–661

Additional Examples

Example 1

Find the opposite of each polynomial.

A. $8x^3y^4z^2$

$-(8x^3y^4z^2)$

The opposite of a is ______.

B. $-3x^4 + 8x^2$

$-(-3x^4 + 8x^2)$

Distribute the sign.

Example 2

Subtract.

A. $(5x^2 + 2x - 3) - (3x^2 + 8x - 4)$

$= (5x^2 + 2x - 3) \quad (\quad 3x^2 \quad 8x \quad 4)$ Add the opposite.

$= 5x^2 + 2x - 3 - 3x^2 - 8x + 4$ Property.

$=$ Combine like terms.

B. $(b^2 + 4b - 1) - (7b^2 - b - 1)$

$= (b^2 + 4b - 1) \quad (\quad 7b^2 \quad b \quad 1)$ Add the opposite.

$= b^2 + 4b - 1 - 7b^2 + b + 1$ Property.

$=$ Combine like terms.

Example 3

Subtract.

A. $(2n^2 - 4n + 9) - (6n^2 - 7n + 5)$

$$\begin{array}{lcl} (2n^2 - 4n + 9) & & 2n^2 - 4n + 9 \\ \underline{-\ (6n^2 - 7n + 5)} & \rightarrow & \underline{+\ -6n^2 + 7n - 5} \end{array}$$

Add the opposite.

B. $(10x^2 + 2x - 7) - (x^2 + 5x + 1)$

$$\begin{array}{lcl} (10x^2 + 2x - 7) & & 10x^2 + 2x - 7 \\ \underline{-\ (x^2 + 5x + 1)} & \rightarrow & \underline{+\ -x^2 - 5x - 1} \end{array}$$

Add the opposite.

Try This

1. Find the opposite of the polynomial.

$-4a^2 + 4a^4$

2. Subtract.

$(c^3 + 2c^2 + 3) - (4c^3 - c^2 - 1)$

3. Subtract.

$(4r^3 + 4r + 6) - (6r^3 + 3r + 3)$

LESSON **13-5**

Multiplying Polynomials by Monomials

pp. 664–665

Additional Examples

Example 1

Multiply.

A. $(2x^3y^2)(6x^5y^3)$

$(2x^3y^2)(6x^5y^3)$

coefficients and

exponents.

B. $(9a^5b^7)(-2a^4b^3)$

$(9a^5b^7)(-2a^4b^3)$

Multiply and add

.

Example 2

Multiply.

A. $3m(5m^2 + 2m)$

$3m(5m^2 + 2m)$ Multiply each term in parentheses by .

B. $-6x^2y^3(5xy^4 + 3x^4)$

$-6x^2y^3(5xy^4 + 3x^4)$ Multiply each term in parentheses by .

Example 3

PROBLEM SOLVING APPLICATION

The length of a picture in a frame is 8 in. less than three times its width. Find the length and width if the area is 60 in^2.

1. Understand the Problem

If the width of the frame is w and the length is $3w - 8$, then the area is ______ or length times width. The answer will be a value of w that makes the area of the frame equal to ____ in^2.

2. Make a Plan

You can make a table of values for the polynomial to try to find the value of a w. Use the ______ Property to write the expression $w(3w - 8)$ another way. Use substitution to complete the table.

3. Solve

$w(3w - 8) =$ ____ $w^2 -$ ____ w ______ Property

w	3	4	5	6
$3w^2 - 8w$				

The width should be ____ in. and the length should be ____ in.

4. Look Back

If the width is ____ inches and the length is ____ times that minus 8, or 10 inches, then the area would be $6 \cdot 10 =$ ____ in^2. The answer is reasonable.

Try This

1. Multiply.

$(5r^4s^3)(3r^3s^2)$

2. Multiply.

$-3a^3b^2(4ab^3 + 4a^2)$

3. Problem Solving Application

The height of a triangle is twice its base. Find the base and the height if the area is 144 in^2.

1. Understand the Problem

The formula for the ____________ of a triangle is one-half base times height. Since the base b is equal to 2 times height, $h = 2b$. Thus the area would be $\frac{1}{2}b(2b)$. The answer will be a value of b that makes the area equal to

____________ in^2.

2. Make a Plan

You can make a table of values for the polynomial to find the value of b. Write the expression $\frac{1}{2}b(2b)$ another way. Use substitution to complete the table.

3. Solve

4. Look Back

If the height is twice the base, and the base is 12 in., the height would be 24 in. The area would be $\frac{1}{2} \cdot 12 \cdot 24 = 144$ in^2. The answer is reasonable.

LESSON 13-6

Multiplying Binomials

pp. 670–671

Vocabulary

FOIL (p. 670)

Additional Examples

Example 1

Multiply.

A. $(n - 2)(m - 8)$

$(n - 2)(m - 8)$ FOIL

B. $(x + 3)(x + z)$

$(x + 3)(x + z)$ FOIL

Example 2

Find the area of the border of a computer screen of width x centimeters around a 50 cm by 80 cm screen. Represent the area of the border in terms of x.

Base: $80 + 2x$ Height: $50 + 2x$

Area of the screen and border combined:

$A = (80 + 2x)(50 + 2x)$

$= \_\_\_ + \_\_\_x + \_\_\_x + \_\_\_x^2$

$= \_\_\_ + \_\_\_x + \_\_\_x^2$

The screen area is ___ • ___ = ___ cm^2, so the frame area is $4000 + 260x + 4x^2 - 4000 = \_\_\_x + \_\_\_x^2$ cm^2.

Example 3

Multiply.

A. $(x + 6)^2$

$(x + 6)(x + 6)$

$+ \quad + \quad +$

B. $(n - m)^2$

$(n - m)(n - m)$

$- \quad - \quad +$

C. $(x - 7)(x + 7)$

$+ \quad - \quad -$

Try This

1. Multiply.

$(r - 4)(s - 6)$

2. Multiply.

$(r - 3)^2$

Chapter 13
Picture Cube

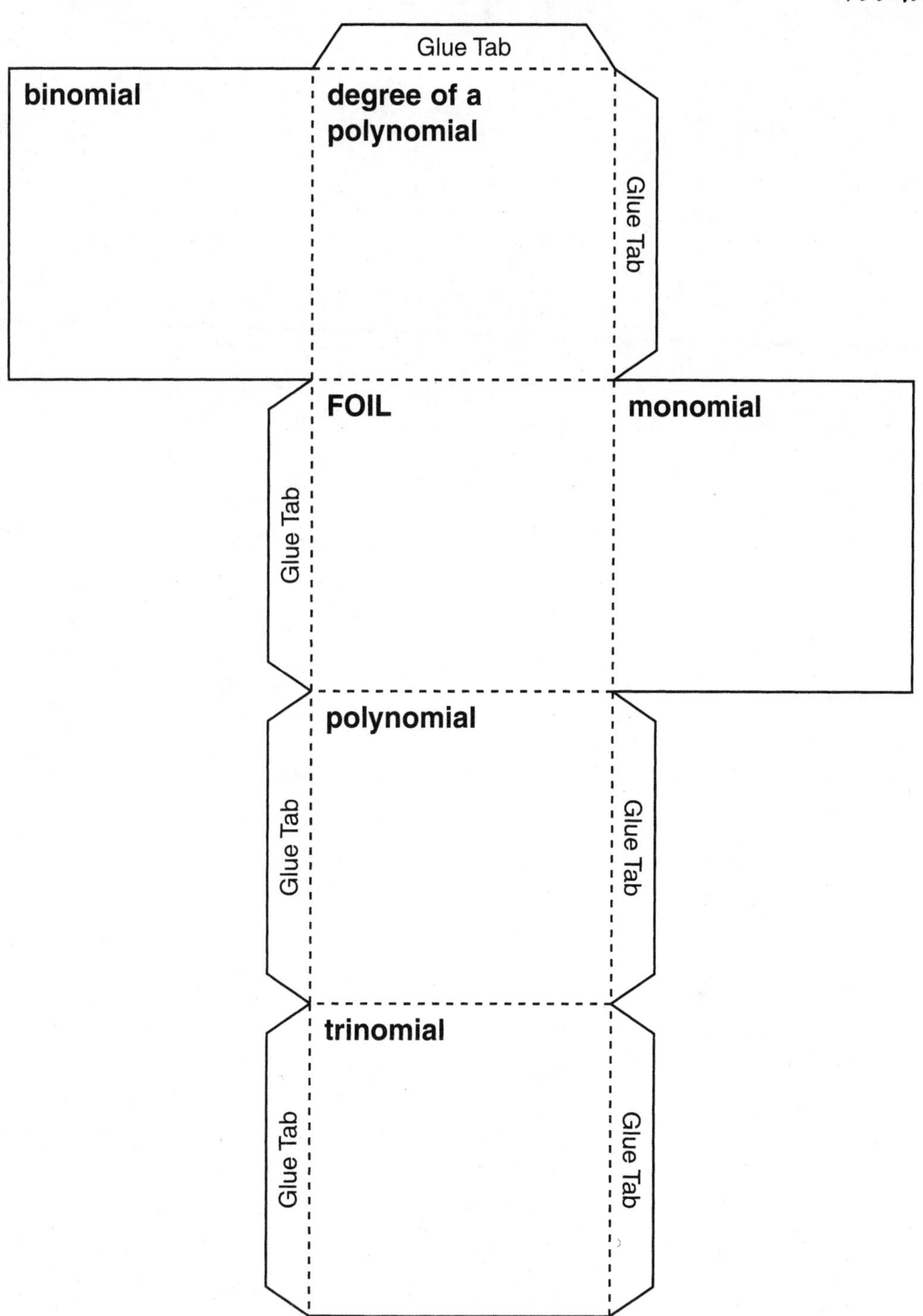

Directions

1. Draw an example of each term on the net of the cube.
2. Cut out the net.
3. Fold along all dotted lines, and place glue tabs to the inside of the cube.
4. Join the common edges, and tape or glue the tabs in place.

LESSON **14-1**

Sets

pp. 688–689

Vocabulary

set (p. 688) ______________________

elements (p. 688) ______________________

subset (p. 689) ______________________

finite set (p. 689) ______________________

infinite set (p. 689) ______________________

Additional Examples

Example 1

Insert $\in$ or $\notin$ to make each statement true.

A. 4 ▢ {even integers}

4 ____ {even integers} 4 is an ____ integer.

B. shoes ▢ {furniture}

shoes ____ {furniture} Shoes are ____ furniture.

C. rhombus ▢ {quadrilaterals}

rhombus ____ {quadrilaterals} A rhombus is a ____.

Example 2

Determine whether the first set is a subset of the second set. Use the correct symbol.

A. E = {even numbers} Q = {rational numbers}

Yes, E Q. Every even number is a number.

B. A = {1, 2, 3, 4} B = {3,4}

No, A B. and are not in the second set.

Example 3

Tell whether each set is finite or infinite.

A. {students in a school}

There are a specific number of students in a school.

B. {points on a line segment}

Between any two points there is always another

.

Try This

1. Insert $\in$ or $\notin$ to make the statement true.

square {rectangles}

2. Determine whether the first set is a subset of the second set. Use the correct symbol.

B = {1, 2, 5, 7} C = {1, 2, 5, 8}

LESSON **14-2**

Intersection and Union

pp. 692–693

Vocabulary

intersection (p. 692) ______________________

____

empty set (p. 692) ______________________

____

union (p. 693) ______________________

____

Additional Examples

Example 1

Find the intersection of the sets.

A. $M = \{1, 3, 5, 7\}$ $N = \{5, 7, 9, 11\}$

The only elements that appear in both M and N are ____ and ____.

$M \cap N = \{$ ____ $\}$

B. $Q = \{\text{rational numbers}\}$ $R = \{\text{real numbers}\}$

Rational numbers are also ____ numbers.

$Q \cap R = \{$ ____ $\}$

C. $P = \{x \mid x < 4\}$ $T = \{x \mid x > 4\}$

There are no ____ in both P and T.

$P \cap T = \{$ ____ $\}$, or ____

Example 2

Find the union of the sets.

A. $M = \{1, 3, 5, 7\}$ $N = \{5, 7, 9, 11\}$

$M \cup N = \{\qquad\}$

B. $Q = \{\text{rational numbers}\}$ $R = \{\text{real numbers}\}$

Real numbers consist of and

numbers.

$Q \cup R = \{\qquad\}$

C. $P = \{x \mid x < 4\}$ $T = \{x \mid x > 4\}$

All real numbers except .

$P \cup T = \{x \mid x \qquad\}$

D. $E = \{\text{even integers}\}$ $O = \{\text{odd integers}\}$

The union of the two sets contains all .

$E \cup O = \{\qquad\}$

Try This

1. Find the intersection of the sets.

$R = \{\text{irrational set}\}$ $S = \{\text{real numbers}\}$

2. Find the union of the sets.

$X = \{2, 3, 4, 10\}$ $Y = \{3, 7, 9, 11\}$

LESSON **14-3**

Venn Diagrams

pp. 696–697

Additional Examples

Example 1

Draw a Venn diagram to show the relationship between the sets.

A. primary colors: {red, blue, yellow}
colors in the American flag: {red, white, and blue}

To draw the ______ diagram, first determine what is in the intersection of the sets.

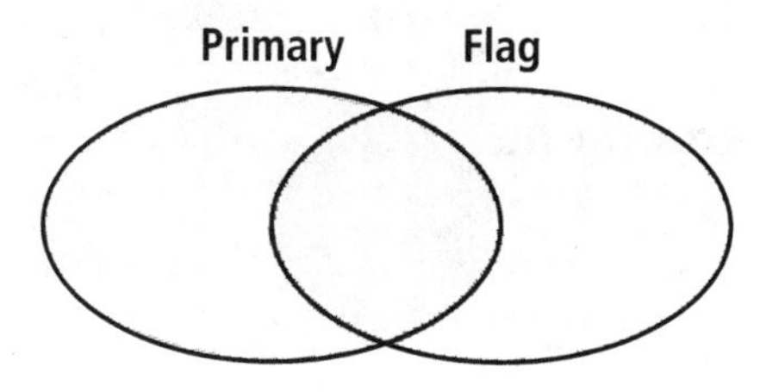

The intersection is { ______ }.

B. Factors of 24 {1, 2, 3, 4, 6, 8, 12, 24}

Factors of 18 {1, 2, 3, 6, 9, 18}

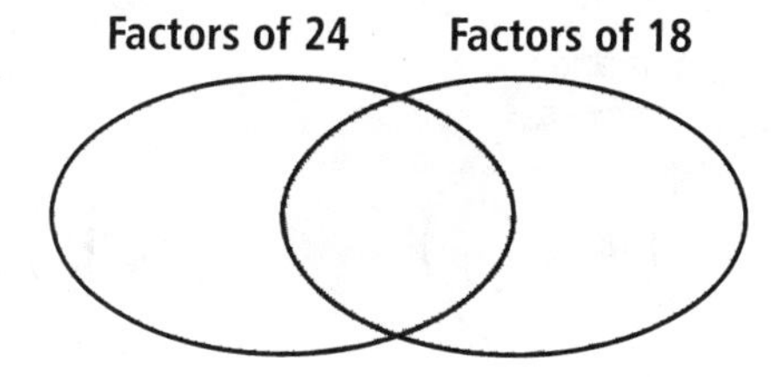

The ______ of the sets is {1, 2, 3, 6}.

Example 2

Use each Venn diagram to identify intersections, unions, and subsets.

A.

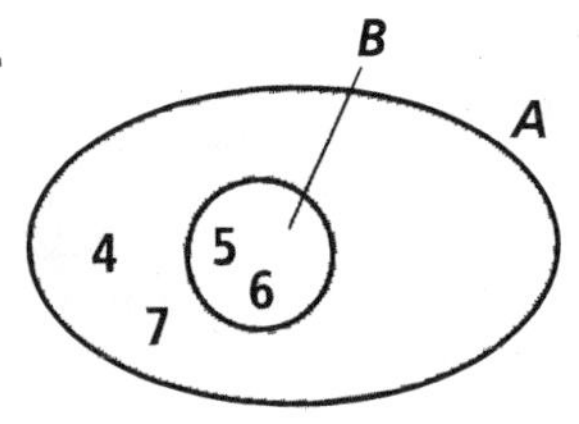

Intersection: $A \cap B = \{ \qquad \}$

Union: $A \cup B = \{ \qquad \}$

Subsets: $\quad \subset$

Example 3

Use a Venn Diagram to show the following logical argument.

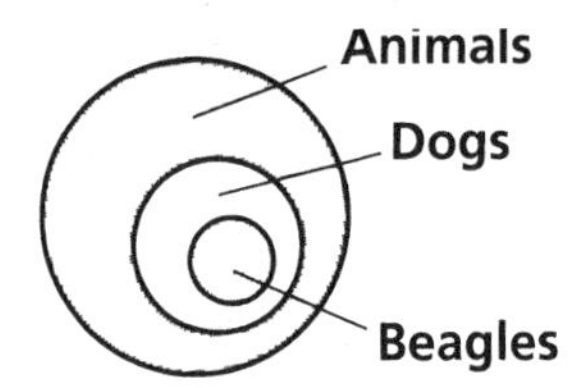

All dogs are animals.
All beagles are dogs.
∴ All beagles are animals.

Try This

1. Draw a Venn diagram to show the relationship between the sets.

letters in the word APPLE {A, P, P, L, E}
letters in the word BANANA {B, A, N, A, N, A}

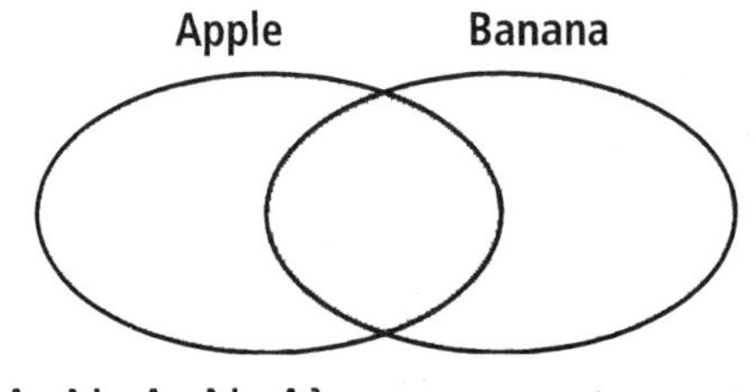

2. Use the Venn diagram to identify intersections, unions, and subsets.

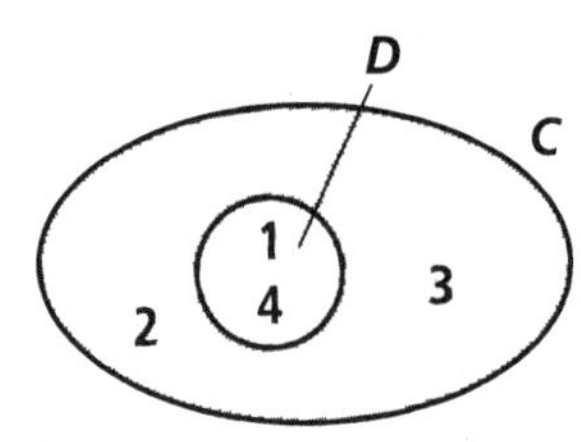

LESSON **14-4**

Compound Statements

pp. 702–703

Vocabulary

compound statement (p. 702)

conjunction (p. 702)

truth value (p. 702)

truth table (p. 702)

disjunction (p. 703)

Additional Examples

Example 1

Make a truth table for the conjunction *P and Q*, where *P* is "Barry is on the soccer team" and *Q* is "Barry plays the cello."

The conjunction *P and Q* is "Barry is on the soccer team ______ he plays the cello."

Example	*P*	*Q*	*P and Q*
Barry is on the soccer team and he plays the cello.			
Barry is on the soccer team and he can't play the cello.			
Barry isn't on the soccer team and he plays the cello.			
Barry isn't on the soccer team and he can't play the cello.			

Example 2

Make a truth table for the disjunction *P or Q*, where *P* is "The sun is shining" and *Q* is "It is a warm day."

The disjunction *P or Q* is "The sun is shining ______ it is a warm day."

Example	*P*	*Q*	*P or Q*
The sun is shining. It is a warm day.			
The sun is shining. It is not a warm day.			
The sun is not shining. It is a warm day.			
The sun is not shining. It is not a warm day.			

Try This

1. Make a truth table for the conjunction *P and Q*, where *P* is "Harry has a high SAT score" and *Q* is "Harry is in the math club."

Example	*P*	*Q*	*P and Q*
Harry has a high SAT score and he is in the math club.			
Harry has a high SAT score and he isn't in the math club.			
Harry doesn't have a high SAT score and he is in the math club.			
Harry doesn't have a high SAT score and he isn't in the math club.			

LESSON **14-5**

Deductive Reasoning

pp. 708–709

Vocabulary

conditional (p. 708)

if-then statement (p. 708)

hypothesis (p. 708)

conclusion (p. 708)

deductive reasoning (p. 709)

premise (p. 709)

Additional Examples

Example 1

Identify the hypothesis and the conclusion in each conditional.

A. If the doorbell rings, then the baby will wake.

Identify the statements following the words ______ and ______.

Hypothesis: ______

Conclusion: ______

Example 2

Make a conclusion, if possible, from the deductive argument.

B. If $x = 5$, then $4x - 1 = 19$.

$x = \sqrt{25}$

Conclusion:

Example 3

Make a conclusion, if possible, from the deductive argument.

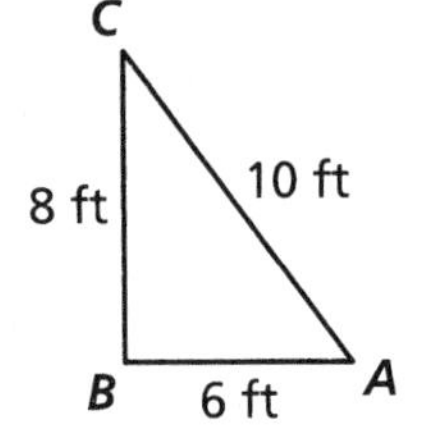

If triangle *ABC* is a right triangle and the legs have lengths of 6 ft and 8 ft, then the hypotenuse has length of 10 ft.

Triangle *ABC* is a right triangle.

$AB = 6$ ft and $BC = 8$ ft.

Conclusion:

Try This

1. Identify the hypothesis and the conclusion in the conditional.

If $x + 2 = 4$ then $x = 2$.

Hypothesis:

Conclusion:

2. Make a conclusion, if possible, from the deductive argument.

If $x = 7$, then $5x + 4 = 39$.
$x = 4^3$.

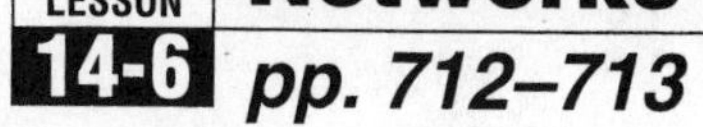

Networks and Euler Circuits

pp. 712–713

Vocabulary

graph (p. 712)

network (p. 712)

vertex (p. 712)

edge (p. 712)

path (p. 712)

connected graph (p. 712)

degree (of a vertex) (p. 712)

circuit (p. 713)

Euler circuit (p. 713)

Additional Examples

Example 1

Find the degree of each vertex, and determine whether the graph is connected.

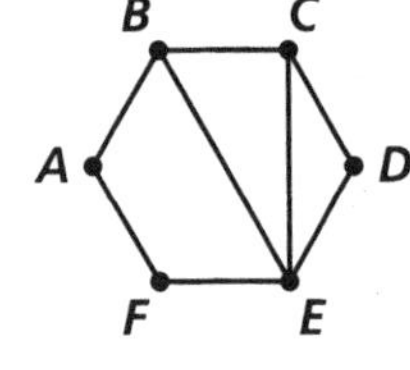

Vertex	Degree
A	
B	
C	
D	
E	
F	

The graph is ______.

There is a path between every ______ and every other ______.

Example 2

Determine whether the graph can be traversed (traveled) through an Euler circuit. Explain.

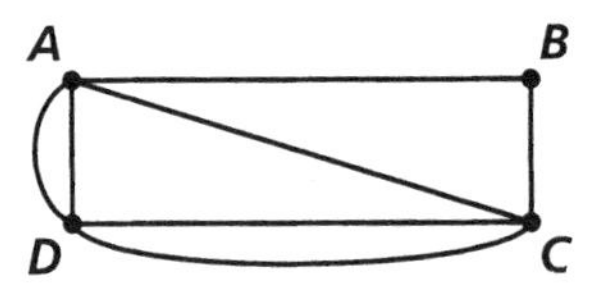

The graph is an ______ circuit because it is connected, and all vertices have ______ degrees.

Try This

1. **Determine whether the graph can be traversed (traveled) through an Euler circuit. Explain.**

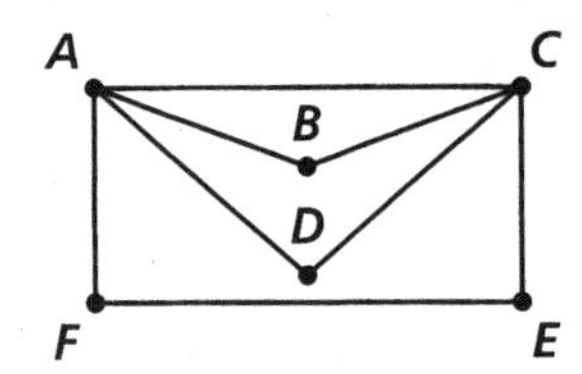

LESSON **14-7**

Hamiltonian Circuits
pp. 716–717

Vocabulary

Hamiltonian circuit (p. 716)

Additional Examples

Example 2

PROBLEM SOLVING APPLICATION

Use the information in the graph to determine the shortest Hamiltonian circuit beginning at Boston.

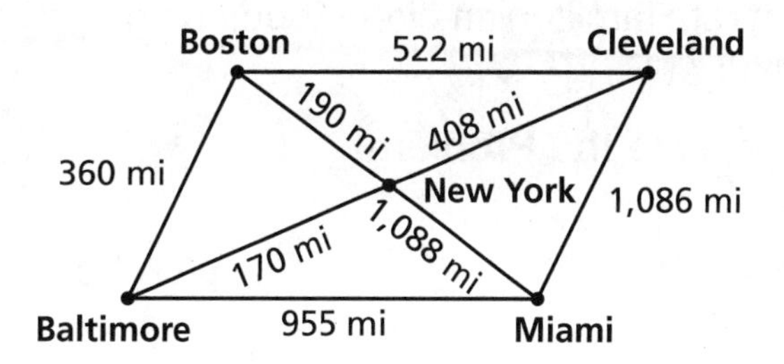

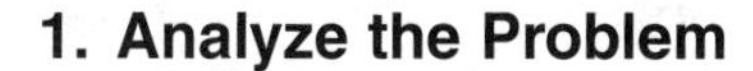

1. Analyze the Problem

Find the   path between the cities.

In the graph the   represent different cities.

2. Choose a Strategy

Find all the Hamiltonian circuits beginning and ending at   .

3. Solve

Find the length of each path. The first letter of each city's name is used to represent the city.

B $\xrightarrow{190}$ N $\xrightarrow{170}$ B $\xrightarrow{955}$ M $\xrightarrow{1086}$ C $\xrightarrow{522}$ B = 2923
B $\xrightarrow{190}$ N $\xrightarrow{408}$ C $\xrightarrow{1086}$ M $\xrightarrow{955}$ B $\xrightarrow{360}$ B = 2999
B $\xrightarrow{522}$ C $\xrightarrow{1086}$ M $\xrightarrow{1088}$ N $\xrightarrow{170}$ B $\xrightarrow{360}$ B = 3226
B $\xrightarrow{360}$ B $\xrightarrow{170}$ N $\xrightarrow{1088}$ M $\xrightarrow{1086}$ C $\xrightarrow{522}$ B = 3226
B $\xrightarrow{522}$ C $\xrightarrow{1086}$ M $\xrightarrow{955}$ B $\xrightarrow{170}$ N $\xrightarrow{190}$ B = 2923
B $\xrightarrow{360}$ B $\xrightarrow{955}$ M $\xrightarrow{1088}$ N $\xrightarrow{408}$ C $\xrightarrow{522}$ B = 3333
B $\xrightarrow{360}$ B $\xrightarrow{955}$ M $\xrightarrow{1086}$ C $\xrightarrow{408}$ N $\xrightarrow{190}$ B = 2999
B $\xrightarrow{522}$ C $\xrightarrow{408}$ N $\xrightarrow{1088}$ M $\xrightarrow{955}$ B $\xrightarrow{360}$ B = 3333

The shortest circuit is   →   ›   –   →   or

        →   ›   –   →   .

4. Look Back

Be sure that you have all possible paths. There are paths that meet the Hamiltonian criteria of only passing through each vertex once.

Try This

1. Problem Solving Application

Use the information in the graph to determine the shortest Hamiltonian circuit beginning and ending at B.

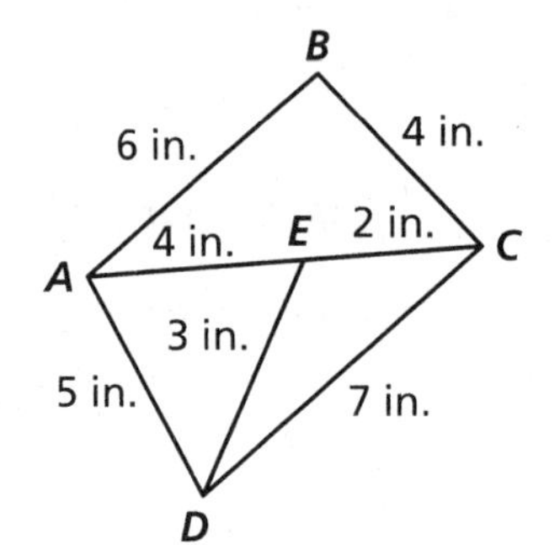

1. Analyze the Problem

Find the path between locations.

In the graph the represent different locations.

2. Choose a Strategy

Find all the circuits beginning and ending at B.

3. Solve

$B \overrightarrow{6}$ $A \overrightarrow{4}$ $E \overrightarrow{3}$ $D \overrightarrow{7}$ $C \overrightarrow{4}$ $B = 24$ in.
$B \overrightarrow{4}$ $C \overrightarrow{2}$ $E \overrightarrow{3}$ $D \overrightarrow{5}$ $A \overrightarrow{6}$ $B = 20$ in.
$B \overrightarrow{6}$ $A \overrightarrow{5}$ $D \overrightarrow{3}$ $E \overrightarrow{2}$ $C \overrightarrow{4}$ $B = 20$ in.
$B \overrightarrow{4}$ $C \overrightarrow{7}$ $D \overrightarrow{3}$ $E \overrightarrow{4}$ $A \overrightarrow{6}$ $B = 24$ in.

4. Look Back

Be sure that you have all possible paths. There are 4 paths that meet the Hamiltonian criteria of only passing through each vertex once and beginning and ending at B.

Chapter 14
Vocabulary Chain

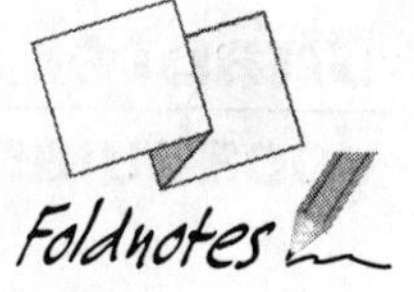

EXAMPLE		TERM
		compound statement

EXAMPLE		TERM
		conjunction

EXAMPLE		TERM
		Euler circuit

EXAMPLE		TERM
		Hamiltonian circuit

EXAMPLE		TERM
		Venn diagram

Directions

1. Write an example and an explanation in words, numbers, and algebra for each term.
2. Cut out each vocabulary strip and fold in thirds.
3. Punch a hole in the corner of each folded vocabulary strip, and string them together to create your vocabulary chain.

Chapter 14
Vocabulary Chain

WORDS	NUMBERS	ALGEBRA
WORDS	NUMBERS	ALGEBRA
WORDS	NUMBERS	ALGEBRA
WORDS	NUMBERS	ALGEBRA
WORDS	NUMBERS	ALGEBRA